Fundamentals of
ICHTHYOLOGY

Fundamentals of ICHTHYOLOGY

by
R.K. Gupta & S.K. Gahlawat
Associate Professor
N.K. Yadava & K.L. Jain
Professor

Department of Zoology and Aquaculture,
CCS Haryana Agricultural University, Hisar 125 004, India

2024

Daya Publishing House®
A Division of
Astral International Pvt. Ltd.
New Delhi – 110 002

First Published, 2007
Reprinted, 2024
ISBN: 9788170354451

Published by : **Daya Publishing House®**
A Division of
Astral International Pvt. Ltd.
– ISO 9001:2015 Certified Company –
4736/23, Ansari Road, Darya Ganj
New Delhi-110 002
Ph. 011-43549197, 23278134
E-mail: info@astralint.com
Website: www.astralint.com

Digitally Printed at : **Replika Press Pvt. Ltd.**

Foreword

India has made substantial advancement in fisheries sector by taking the advantage of physical and climatic conditions. More than 3000 crore rupees worth of fish and fish products are being exported to other countries generating a handsome amount of foreign exchange.

The book entitled *"Fundamentals of Ichthyology"* prepared by our faculty is highly useful book which may serve the long felt needs of students, teachers and research scholars. I believe that this work will be of great use for fisheries under graduate and post-graduate students. Initial chapters deals with some basic aspects of culturable fishes whereas the middle segment of this book deals with nutritional aspects of fish culture and lastly this book through light on the various aspects of diseases management. I congratulate our faculty for their valuable efforts in bringing out this book.

R.C. Sihag
Dean
College of Basic Sciences and Humanities
CCS Haryana Agricultural University, Hisar

Preface

This book entitled *"Fundamentals of Ichthyology"* has been written compiling information to meet the needs of fisheries student at graduate as well as post-graduate level. The prime objective of the book is that the student should learn about identification description of fishes and the various basic aspects about fish culture. The first three chapters describes about the various culturable varieties of fishes. Then we gradually move to chapters on feeding and supplementary feeds and in the last we described about the various management practices of fish culture particularly disease management. In the last various aspects of fish and prawn diseases were mentioned briefly. These chapters can be useful to create awareness among the persons associated with the aquaculture industries about the havoc of diseases.

We are thankful to the CCS Haryana Agricultural University authorities and the publisher for motivating us to write this book and provide financial assistance for this. We are also indebted to various authors whose valuable works have been cited in this book.

Authors

Contents

Chapter 1

Nature and Life of Fishes

Fish are cold blooded vertebrates that breathe by means of gills and live in water. They comprise about 30,000 to 40,000 species differing widely from each other in shape, size, habits and habitats. Some of them are small in size not more than 1 inch, but some are very large about 15 meter length. They live in the sea, rivers, lakes, canals, dams and in almost every place where there is water. Fishes are streamlined body but some are elongated snake-like and a few are dorso-ventrally compressed. They have paired and unpaired fins supported by soft or spiny rays. Dorsal, caudal and anal fins are unpaired while the pectoral and the pelvic fins are paired. A number of species possess barbells which are an excellent organ of touch. Besides being used as food, fish liver is an important source of oil containing vitamin A & D. Body oil from fish is extensively used in soap industries and tanning. As compared to tetrapod, the fishes have the following common characters, most of which are due to their permanent aquatic life:

1. Gills are the respiratory organs.
2. Locomotion by paired or unpaired fins.
3. Paired visceral arches are present.
4. An exoskeleton of scales is usually present.
5. Kidney is mesonephric.
6. A swim bladder is usually present.
7. The notochord is partly replaced by cartilaginous or bony vertebrae.
8. There is no middle ear.

Classification

The classification of fish means the sorting of each kind of individuals into groups. In the eighteenth century basis of classification was on the existence of natural affinities. Each group was arranged according to their natural affinities which were combined into large categories and so on. J. Muller (1844) gave the first scientific classification of the lower vertebrates and divided pisces into six sub classes:

1. Dinpoi
2. Teleostei
3. Ganoidei
4. Elasmobranchii
5. Cyclostomi
6. Leptocardii

The modern ichthyologists have tried to classify them into super classes, classes, superorders, orders, divisions, suborders, families, sub-families, genera etc. in the following account broad classification of living fishes and principal kind of fishes are included. These are aquatic gastrostromes having gills as the respiratory organs and fins for locomotion. These are divided into seven classes:

1. Pterichthys
2. Cocostei
3. Acantrodii
4. Elasmobranchii
5. Holocephali
6. Dipnoi
7. Telestomi

Of these, the first three classes are completely extinct and are exclusively known as placoderms.

The most important aspect of fish biology is the mode of their propagation in natural water. The act of reproduction varies widely among different species. In the act of copulation male gentalia are inserted into this cloacae of the female and serve as the passage for the semens. In this way ova are fertilized internally. This means the union of the ova and sperms inside the body of the female in water. The fishes have chosen to produce very large number of small eggs running into lacs. Most of them are lost or devoured, but a large number live to go through the larval and young stages and grow up to reproduce their own kind externally .

Many fish species undertake long journeys either within the same ecological niche or from their original habitats to another. Myer (1949) has proposed the following terms to describe the fish migration.

1. Diadromous fish ← Anadromous / Amphidromous / Catadromous
2. Potamodiomous
3. Oceamodiomous

Diadromous Fishes

These are truely migratory fish which migrate between the sea and freshwater, and are of three types:

Anadromous

Diadromous fishes which spend a major part of their live in a sea, migrate to freshwater during breeding period for spawning, *e.g.*, many marine fishes like the salmon, sea lamprey, piles etc.

Catadromous

This group includes diadromous fishes which spend a major part of their live in freshwater but migrate to sea for breeding purpose, *e.g.*, the freshwater Eel, Anguilla etc.

Amphidromous

These are diadromous fishes in which migration from freshwater to the sea or vice versa, is not a purpose of breeding but occurs regularly at some other definite stage of the life cycle. Myer suggests that the migration of some gobies might fall in this category.

Potamodiomous

Truly migratory fish whose migration remains confined to freshwater, *e.g.* the carps and the trout travel long distances in large rivers in search of spawning grounds. After egg laying at suitable places, they return to the feeding area.

Oceamodromous

Truly migratory fish which live and migrate in the sea. Many marine fishes like the herrings (Clupea), macherels (Scomber) travel for long distances in the sea to deposit their eggs and later return to the feeding grounds.

The Mahaseer (*Tor* spp.) undertake long distance migration from the mediterrian rivers of the peltries to mountains for breeding, coming down with the lowering of temperature after post-monsoon in the hills. The foregoing account summarizes broadly the nature and life of fish highlighting the diversity of form and their mode of living. The reader, if interested may see books available on the life and history of fishes.

Chapter 2

General Classification and Measurement of External Parts of the Body of a Fish

Definition of Fish

Fishes are cold blooded aquatic vertebrates which breathe by means of pharyngeal gills, propelling and balancing themselves by the means of fins. Some 20,000 species are known to inhabit water bodies of the various descriptions.

General Classifications of Fishes

The living forms included in superclass Pisces are divided into four classes, *viz.* Elasmobranchii, Holocephali, Dipnoi and Teleostomi, of which, Dipnoi does not occur in India. The diagnostic characteristics of Indian representatives of the three classes are:

Class-Elasmobranchii

Skeleton cartilagenous; skin of body covered with placoid scales or naked; 5-7 gill openings on each side; no gill cover, teeth distinct; cloaca present; males with claspers, *e.g.* Rays and Sting Rays

Class-Holocephali

Skeleton cartilagenous, skin of the body covered with placoid scales; 5 gill arches; 4 gill openings protected by a gill cover or the operculum; teeth united to form grinding plates; no cloaca; males with claspers, *e.g.* Chimarea.

Class-Dipnoi

The Dipnoi (double breathers) is an important group of bony fishes. They are also called as "Lung fishes" and represent transitional group between fishes and amphibians. Class-Dipnoi comprises only two orders: Monopneumona (with single lung), and Dipneumona (with double lungs). The fishes of this class are having long fusiform body, strong heterocercal tail, archipterygial type paired fins, two dorsal fins and large heavy cosmoid scales, ossified jaws and dentitions highly specialized with suppressed marginal teeth in both the jaws. These fishes frequently come to surface and breathe air.

Class-Teleostomi

Skeleton bony; skin of the body covered with ganoid, ctenoid scales or naked; 5 gill arches; 4 gills; a pair of lateral gill-opening confluent as two lateral slits with operaculum; usually no cloaca. Almost all freshwater fish are covered under this class.

Different Parts of the Body of Fish (Figure 1)

1. Eye (without eye lids)
2. External nostrils

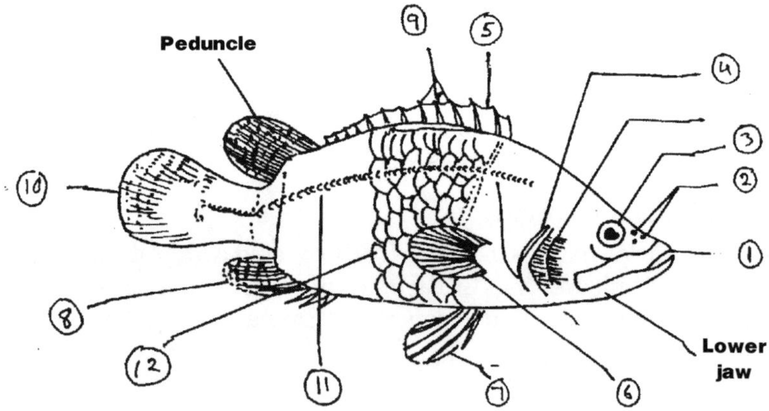

Figure 1: Body Parts of a Fish

(1) Mouth; (2) External nostrils; (3) Eye; (4) Operculum; (5) Dorsal fin; (6) Pectoral fin; (7) Ventral fin; (8) Anal fin; (9) Fin rays; (10) Caudal fin; (11) Lateral line; (12) Scale

3. Mouth consists of Pre-maxilla Maxilla, Mandibles, Teeth Grinding plates, Lip, Mouth cavity and Mouth tentacles
4. Operculum consists of pre-operculum, inter-operculum and sub-operculum
5. Fins are Dorsal, Ventral, Pectoral, Anal and Caudal
6. Fin ray spines (Dorsal, Pectoral, Anal and Ventral)
7. Additional Dorsal Fin or Peduncle
8. Scales (Placoid, Ganoid, Cycloid and Ctenoid)
9. Lateral line (LI)
10. Latero-transversal line (Ltr).

Measurement of Different Parts of the Body of a Fish

Name of the Organ	*Way of measurement*
Eye	By taking total diameter of eye
Snout	By taking length between mouth and eye
Length of head	By measuring distance between mouth and postoperculum
Length of body	By measuring distance between mouth and commencement of caudal fin
Total length	By measuring vertical distance between mouth to end of caudal fin
Depth of body	By measuring vertical distance between peduncle (dorsal fin) and anal fin

Fish Formula (FF)

It may be defined as the formula by which the exact number and measurements of different parts of the body/number of fin rays of different fins and number of scales present on the lateral line as well as on the latero-transverse line may be counted. It is the key to identify and differentiate among species of the same genera, *e.g.*

$$FF = D_{14\text{-}16}, A8, V_9, P_{21}, C_{19}, LI_{40\text{-}42}, Ltr_{7.5\text{-}9}$$

where,

D = No. of fin-rays of Dorsal fin,

A = No. of fin-rays of Anal fin,

V = No. of fin-rays of Ventral fin,

P = No. of fin-rays of Pectoral fin,

C = No. of fin-rays of Caudal fin,

Ll = No. of scales present on the lateral line of the body of fish

Ltr = No. of scales present on the latero-transverse line of the body of fish.

Chapter 3

Classification, Habits and Habitat of Some Important Culturable Freshwater Fishes

The class Teleostomi commonly known as teleosts includes families Cyprinidae, Siluridae and Ophiocephalidae. The general description of these families is as follows:

Family: Cyprinidae

Body oblong or elongated, abdomen usually rounded, but may be compressed, margin of upper jaw is formed by the pre-maxillaries

Operculum is in four pieces. Mouth toothless, but sometimes form one to three rows of teeth in the interior pharyngeal bones

A single rayed dorsal fin

Head scaleless. Body scaled but rarely scaleless.

Mostly herbivorous in food habits.

Family: Siluridae

Margin of the upper jaw formed mainly by pre-maxillaries, Maxilla is rudimentary and often constitutes the base of a barbel

Barbels are present

No sub-opercle

Either rayed dorsal fin or adipose dorsal fin, may be present or absent

Skin scaleless

Mostly carnivorous in food habits

Mouth is armed with teeth.

Family: Ophiocephalidae

Body elongated, sub-cylindrical anteriorly, head depressed having superiorly somewhat plate like scales

Eyes are lateral. Gill opening wide

Teeth are present in the jaws or vomer or plates, some of which may be conical

A single, long, spineless dorsal fin

Ventral fin, thoracic fin may be present or absent, when present, consisting of six rays

Scales are of large, moderate or small in size

Carnivorous in food habits.

Identification of Some Economically Important Cultivable Fishes

1. Family: Cyprinidae

(a) Major Carps

(*i*) *Catla catla*

(*ii*) *Labeo rohita*

(*iii*) *Cirrhina mrigala*

(b) Minor Carps

(*i*) *Labeo calbasu*

(*ii*) *Cirrhina reba*

(*iii*) *Barbus sarana*

(c) Exotic Carps

(*i*) *Ctenophryngodon idella*

(*ii*) *Hypopthalmicthys molitrix*

(d) Common Carp

(*i*) *Cyprinus carpio* var. communis

(*ii*) *Cyprinus carpio* var. specularis

(*iii*) *Cyprinus carpio* var. nudus

2. Family: Siluridae

(a) Cat Fishes

(*i*) *Wallago attu*

(*ii*) *Hateropnuestes fossilis*

(*iii*) *Clarias magur*

(*iv*) *Mystus seenghala*

3. Family: Ophiocephalidae

(b) Murrel Fishes

(*i*) *Ophiocephalus striatus*

(*ii*) *Ophiocephalus punctatus*

(*iii*) *Ophiocephalus marulius*

(*iv*) *Ophiocephalus gachua*

Indian Major carps

(a) Catla *(Catla catla)*

Class : Teleostomi

Order : Physostomi

Family : Cyprinidae

Genus : *Catla*

Species : *catla*

It is commonly called catla and is distributed in almost all States of India; though, its natural distribution extends only to the Godavari River in Andhra Pradesh, but now distributed in almost all the river systems.

It is the fastest growing carp in India.

Characteristically, bulged body with a conspicuous head, large upturned mouth, non fringed lips, devoid of barbels and a dorsal fin with 14 to 16 branch rays distinguish the adult fish and advanced fingerlings.

The body is ordinarily dull, silver white, but tends to be rather darkish in weedy waters.

The fry, from half to one inch in size, are distinguished by the relatively large head with reddish gills.

Catla is a surface and column feeder and its upturned mouth and large gill rakers are adapted to feed on numerous organisms, floating in water.

Submerged weeds are of no use to catla and in ponds with an overgrowth of weeds, its growth is generally unsatisfactory.

It is reported to grow very quick even unto three to four inches per month.

In normal stocked waters, a growth of 15 to 18 inches in the first year can usually be expected.

In lifetime, catla grows to over four feet in length.

Fish in the second year, are ordinarily sexually ripe. Specimens 22 months old measuring 18 inches in total length and weighing about 3 pounds, have been observed sexually mature.

Catla breeds in river during monsoon months from June to August.

Usually it does not breed in pond or lakes especially in captivity.

FF = D.14-16, P.21, V.9, A.8 (3/8), C.19, L1. 40-42, Ltr. 7½/9

(b) Rohu (*Labeo rohita*)

Class : Teleostomi

Order : Physostomi

Family : Cyprinidae

Genus : *Labeo*

Species : *rohita*

A quick growing carp, with almost the same distribution as catla and considered a tastier fish than any other Indian carp.

It is commonly known as rohu, distributed to Bengal, Bihar, Assam, Uttar Pradesh, Punjab, Orissa and Andhra Pradesh.

Rohu is easily distinguished by its small or pointed head, almost terminal mouth with fringed lower lip, dull reddish scales on the sides and pink reddish fins.

The dorsal fin has 12 to 13 branched rays.

The body is more linear than that of catla.

Rohu is a column-bottom feeder and its rather anterior fringe-lipped mouth is well adapted to browsing in the shallow pond's bottom also. Considerable quantities of bottom sand or mud, vegetable debris or decaying leaves of aquatic plants, planktonic algae, etc. constitute the stomach contents.

Water fleas or micro animalcules are of no use as food.

Rohu grows very quick, though perhaps, relatively a little slower than catla.

A growth of 14 to 16 inches can normally be expected in the first year in a well stocked fish pond. Sexual maturity is attained towards the end of the second year.

Rohu grows to over three feet in length.

Like catla, rohu also breeds in rivers and also in the bund type tanks, during the monsoon and its frys are collected from these habitats for rearing and stocking.

FF=D.15-16, P.17.V.9.A.7(2/5), C.19. LI.40-42, Ltr. 7½/9

(c) Mrigal (*Cirrhiana mrigala)*

Class : Teleostomi

Order : Physostomi

Family : Cyprindae

Genus : *Cirrhina*

Species : *mrigala*

Next in importance to catla and rohu for culture purposes is the mrigal.

Common in major river systems including river Godavari.

Locally called mori, distributed mainly in states of Bihar, Bengal, Orissa, Uttar Pradesh, Punjab and Andhra Pradesh.

It is easily distinguished by the relatively linear body, small head with rather blunt snout, terminal mouth with thin non-fringed lips, bright silvery body and reddish fins.

The dorsal fin has 12-13 branched rays. The young ones can easily be distinguished by their blunt snout and thin lips.

In slight larger specimens, the scales at the side of the body have their edges pigmented dark.

The fins also become reddish.

Mrigal is a bottom-column feeder.

The fingerlings and adults take relatively large quantities of decaying vegetable and organic debris, phyto-planktons, sand and mud, as compared to rohu.

The proportion of animal matter in the diet is generally poor.

The thin terminal lips are adapted to picking up things from the mud and from the invariable presence of appreciable quantities of sand and mud in the gut, algae in the fresh or decaying conditions are also picked up from the bottom.

Mrigal grows slower than catla and rohu.

The species attain a length of over three feet.

Mrigal becomes sexually mature only when about two years old.

The breeding season and source of fry are the same as those of catla and rohu.

FF=D.15-16, P.15, V.9, A.8 (3/5), C.15, L1.40-45, Ltr. 6½-7/8½

Minor Carps (Family: Cyprinidae) (Figure 2)

(a) Kalbasu (*Labeo calbasu*)

Class	:	Teleostomi
Order	:	Physostomi
Family	:	Cyprinidae
Genus	:	*Labeo*
Species	:	*calbasu*

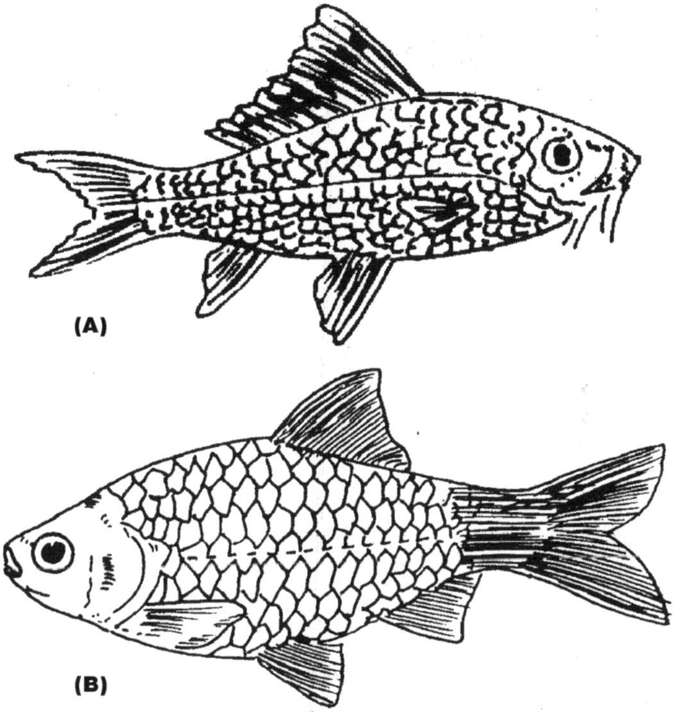

Figure 2: Indian Minor Carps
(A) *Labeo calbasu*; (B) *Barbus sarnana*

Kalbasu occurs almost throughout India, except in Malabar.

It does not generally occur in large numbers, but whenever available, it is stocked in ponds.

Though attaining a maximum length of 2.5 feet or more, it is relatively slow growing fish.

The conspicuous, uniformly darker (blackish gray) colour of the body, the small tapering head with the sub-terminal fringed-lipped mouth, and conspicuous barbels are its distinguishing characters, the dorsal fin has 12 to 13 branched ray in rohu and mrigal.

The fry are easily distinguished by their peculiar pigmentation.

Kalbasu feeds on snails and worms at the bottom of the pond and also on algae and other decayed matter, but the animal portion of the diet is generally very small.

In well stocked ponds, an average annual growth of 10-12 inches may be expected in the first year. Often small specimens, about 8.5 to 9 inches long, are sexually ripe; there is no direct evidence that they attain maturity in the first year. Usually in the large specimens, maturity is attained in the second year only.

FF=D.16-18 P.19, V.9, A.7(2/5), C.19, LI.40-44, Ltr. 7½/8

(b) Reba (*Cirrhina reba*)

Class	:	Teleostomi
Order	:	Physostomi
Family	:	Cyprinidae
Genus	:	*Cirrihina*
Species	:	*reba*

It is a widely distributed minor carp, but has not so far been recorded in Malabar.

Though, its maximum recorded size is 12 inches, it is extensively used for stocking purpose in Madras.

The characteristics hexagonal scales on the body, the lateral dark pigment band on either side and the small tapering head are its distinguishing features.

The dorsal fin has only eight to nine branched rays.

A dark pigmented band, originating on the snout and extending up to the tail on either side, is also a distinguishing character.

Reba appears to subsist largely on planktonic algae which form about sixty per cent of the food.

Considerable quantities of mud, sand and debris are also met within the gut contents.

The young ones feed voraciously on zooplanktons and sometimes grow faster than even the young ones of catla.

Reba has more pronounced plankton feeding habits.

Though a minor carp, *Cirrhina reba* grows fairly quickly, and by the end of first year of life, attains a length of about 11 inches.

It breeds in rivers during monsoon months. Carp fry are collected from reverine habitats.

FF=D.10-11, P.16, V.9, A.8(3/5), C.19, LI. 35-48, Ltr. 7/9

(c) Sarana (*Barbus sarana*)

Class : Teleostomi

Order : Physostomi

Family : Cyprinidae

Genus : *Barbus*

Species : *sarana*

This is medium sized fish and attains a length of over a foot and distributed throughout India.

Though, there is no regular culture of this species in certain localities, particularly in Chennai (Madras), fingerling are collected for stocking in ponds.

It is a voracious feeder, consuming quantities of decaying macro-vegetations, gastropod shells, worms, insects and filamentous algae.

It breeds in rivers during monsoon months but not in ordinary ponds.

The young ones are caught along with the fry of major carps and stocked in ponds.

Usually only a few specimens survive and grow rapidly, attaining a length of about 10 inches in the first 10 months of their life, and sexual maturity also at the end of this period.

FF=D.11, P.15, V.9, a.8 (3/5), C.19, L1. 32-34, Ltr. 5½ -6/6

Cat Fishes (Family: Siluridae) (Figure 3)

(a) Freshwater Shark (*Wallago attu*)

Class : Teleostomi

Order : Physostomi

Family : Siluridae

Genus : *Wallago*

Species : *attu*

One of the largest of the freshwater catfishes attaining a length of about six feet, *W. attu* is widely distributed throughout India.

It is highly predatory in nature and highly destructive to carps. Its large mouth, well armed with teeth, enables it to successfully attack and destroy even medium sized carps.

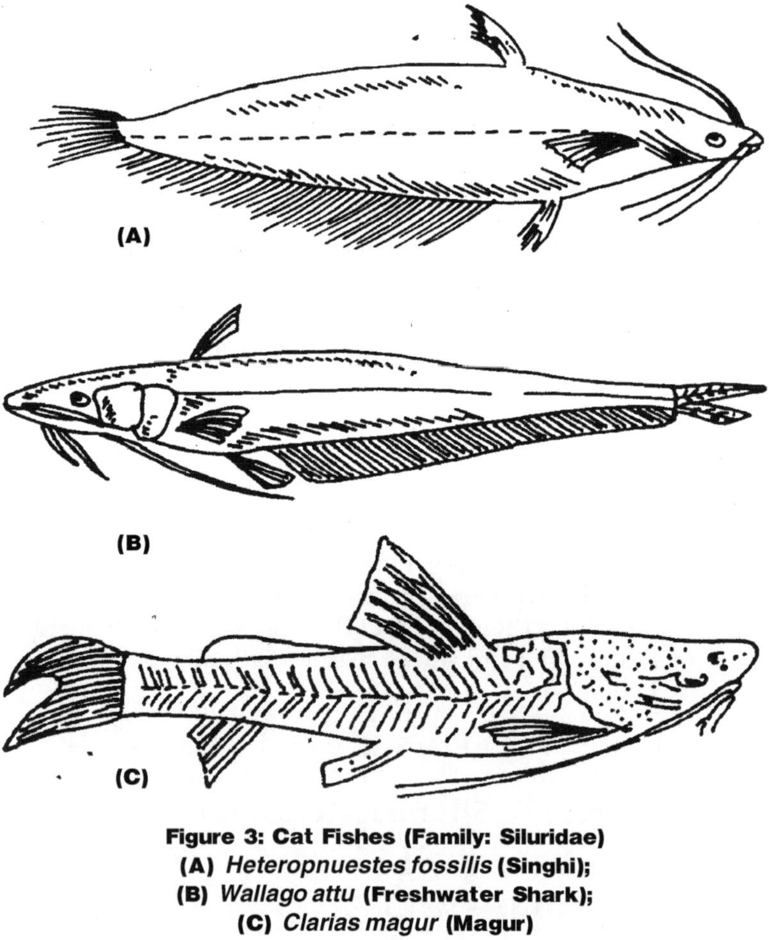

(A)

(B)

(C)

Figure 3: Cat Fishes (Family: Siluridae)
(A) ***Heteropnuestes fossilis*** **(Singhi);**
(B) ***Wallago attu*** **(Freshwater Shark);**
(C) ***Clarias magur*** **(Magur)**

A variety of items including limbs have been observed in its stomach.

It breeds in rivers during the rainy months.

It is reported to attain a length of about 30 inches in the first year of its life.

On account of its predacious habits, this species should be carefully removed from carp ponds.

(b) Singhi (*Heteropnuestes fossilis*)

Class	:	Teleostomi
Order	:	Physostomi
Family	:	Siluridae
Genus	:	*Heteropnuestes*
Species	:	*fossilis*

This is a common catfish with accessory respiratory organs and therefore, able to breath atmospheric oxygen.

It is available throughout India and attains a length of over a foot.

Though not as highly priced food fish as Murrel, its flesh is supposed to have invigorating qualities.

The fish is very much feared for its poisonous pectoral spines which can inflict very painful wounds leading to local inflammation and even fever.

It is predacious in nature but not markedly piscivorous. Insects, ostracodes, worms, algal matter, organic debris and fish and fish remains constitute the main items of food.

Varying quantities of mud or sand are also often taken in. It breeds during April-June but very little is known of its history/growth.

(c) Magur (*Clarias magur*)

Class	:	Teleostomi
Order	:	Physostomi
Family	:	Siluridae
Genus	:	*Clarias*
Species	:	*magur*

This is a catfish with habits very similar to that of singhi but not so common as the latter, though found throughout India and growing to much larger size.

Being an esteemed food fish, it is always in high demand.

Magur is predatory like singhi but not so markedly piscivorous.

Medium sized specimens have been found to take prawns in appreciable numbers, aquatic insects and their larvae, sand and mud occasionally mixed with algal filaments.

The fish breeds in ponds during April-June but in different types of environments.

(d) Seenghara (*Mystus seenghala*)

Class : Teleostomi

Order : Physostomi

Family : Siluridae

Genus : *Mystus*

Species : *seenghala*

This is another large reverine catfish, distributed almost throughout India.

It attains a length of about six feet and is sill in greater demand than *W. attu* as a food fish.

The fish is a predator with marked piscivorous tendencies

Barbus filamentous and *Cirrhina reba* have been found most frequently in its stomach in the Cauvery river.

Insects and their larvae, crustaceans, gastropods and very rarely aquatic weeds are also consumed.

The fish breeds in rivers as well as in ponds during April to July.

Its breeding habits are peculiar because it shows parental care in rearing its eggs and young ones.

Murrel or Live Fishes (Family: Ophiocephalidae) (Figure 4)

(a) Common Murrel (*Ophiocephalus striatus*)

Class	:	Teleostomi
Order	:	Physostomi
Family	:	Ophiocephalidae
Genus	:	*Ophiocephalus*
Species	:	*striatus*

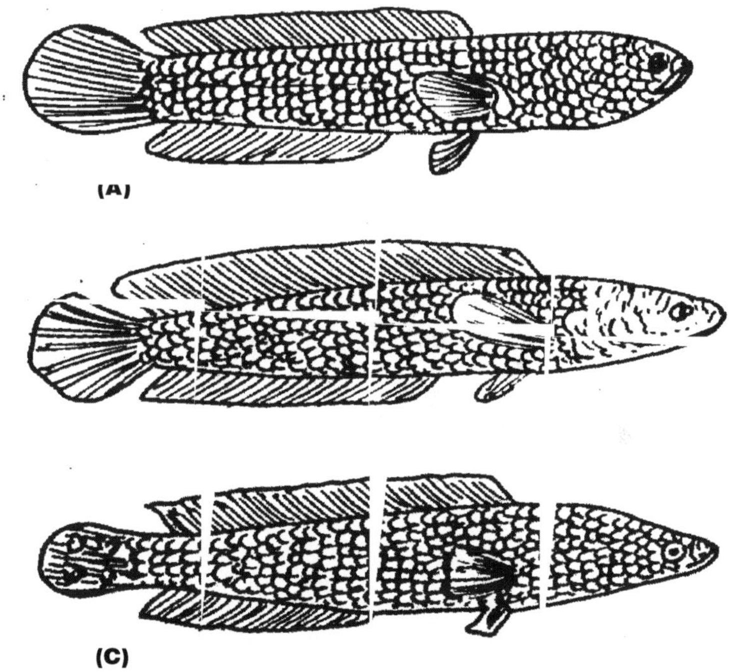

Figure 4: Murrel Fishes (Family: Ophiocephalidae)
(A) ***Ophiocephalus punctatus*** **(Kaveri Murrel);**
(B) ***Ophiocephalus striatus*** **(Common Murrel);**
(C) ***Ophiocephalus marulius*** **(Large Murrel)**

Murrel forms the bulk and mainstay of the pond fisheries in several parts of India.

It is regarded as an excellent table fish in Punjab, Madhya Pradesh, Hyderabad, Mysore and Travancore–Cochin, and is highly priced and is in greater demand than carps.

Ophiocephalus striatus is perhaps the commonest and, economically, the most important food fish.

The different species of murels resemble each other closely.

Being a predator, it feeds on minnows, minor carps, major carps, fingerlings, insects, frogs and other live food, fries feed on zooplanktons and fingerlings feed mainly on insect larvae and young fish, if available.

It breeds in ponds almost throughout the year, with a peak period immediately preceding, and during the monsoon months.

Sexual maturity is attained at a length of about 9.5 or 10 inches.

Eggs are laid in floating mats or scums and parental care is also observed in this fish.

Under favourable conditions, a quick growth of up to 19 inches has also been recorded.

(b) Kaveri Murrel (*Ophiocephalus punctatus*)

Class	:	Teleostomi
Order	:	Physostomi
Family	:	Ophiocephalidae
Genus	:	*Ophiocephalus*
Species	:	*punctatus*

This is one of the small species of the genus attaining a foot in length and subsiding largely on small fishes and aquatic insects.

Breeding season and habits are similar to those of *O. striatus.*

Very little is known about its growth, but sexual maturity is probably attained in the first year of its life as in *O. striatus.*

(c) Large Murrel *(Ophiocephalus marulius)*

Class : Teleostomi

Order : Physostomi

Family : Ophiocephalidae

Genus : *Ophiocephalus*

Species : *marulius*

Probably the largest among the murrels, *O. marulius* attains a length of about four feet and is found almost throughout India.

It prefers large body of water, like lakes, reservoirs and swamps, though it is also found in rivers.

It does not breed in small ponds. It is so highly predacious, even cannibalistic, that when stocked in confined waters like irrigation wells, the survival is limited to only a very few.

It breeds from April to June and breeding habits are similar to those of *O. striatus.*

The young ones grow fast, a maximum growth of about 30 inches at the end of first year of its life is on record.

(d) Small Murrel *(Ophiocephalus gachua)*

Class : Teleostomi

Order : Ophiocephalidae

Family : Ophiocephalus

Genus : *Ophiocephalus*

Species : *gachua*

This is called a minor or small murrel because of its small size, *O. gachua* is of very limited economic significance, it attains a maximum size of four to six inches, feeds on small fries, fingerlings, aquatic insects, micro crustaceans and other small aquatic insects.

Sexual maturity is attained at the end of first year.

Exotic Carps (Family: Cyprinidae) (Figure 5)

Some exotic fishes of sport, food and public health significance have been introduced in India. As these fishes were very hardy so they easily flourished in this country. Following are more important from the point of view of fish culture.

(a) Grass Carp (*Ctenopharyngodon idella*)

Class	:	Teleostomi
Order	:	Physostomi
Family	:	Cyprinidae
Genus	:	*Ctenopharyngodon*
Species	:	*idella*

This fish is the natural inhabitant of the flat land rivers of China and the middle and lower sections of river Amur in USSR.

Grass carp has been introduced in many countries including India from Hong Kong.

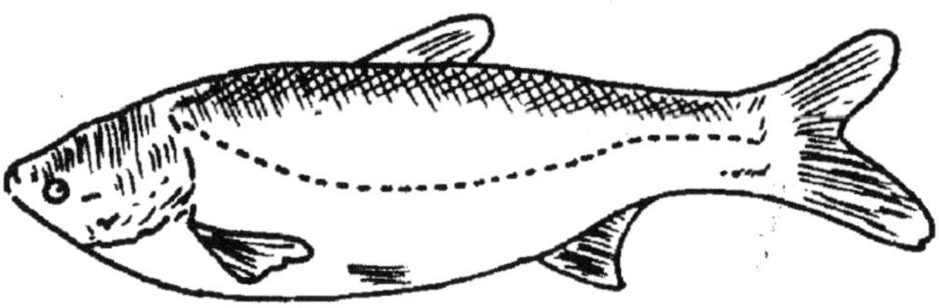

(A)

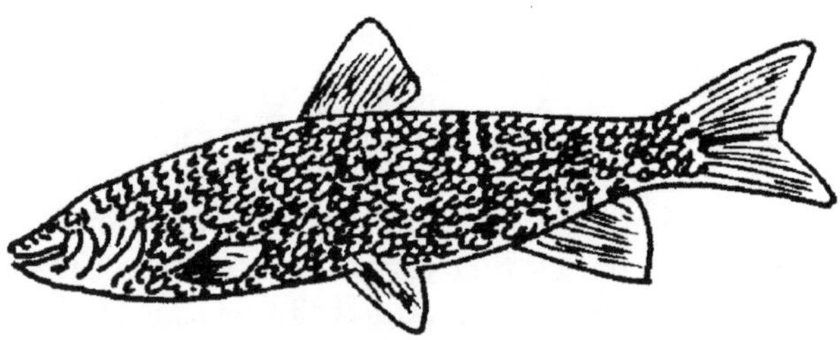

(B)

Figure 5: Exotic Carps (Family: Cyprinidae)
(A) ***Hypopthalmicthys molitrix*** **(Silver Carp);**
(B) ***Ctenopharyngodon idella*** **(Grass Carp)**

Head is broad with short round snout, upper jaw slightly longer than the lower, no barbel, mouth sub-terminal and fin round, colour of the body dark grey above and silvery on the belly, all fins dark, base of each scale dark brown.

The maximum length reported is 120 cm and weight up to 32 kg.

Grass carp, though a freshwater fish, is also able to tolerate slightly brackish water like the silver carp and is almost exclusively a vegetarian.

Like other Cyprinids it has a toothless mouth but has strong and specialized pharyngeal teeth for grasping aquatic vegetations.

These teeth occur in two rows.

The spawning season of grass carp is in summer approaching to monsoon to monsoon season in the flooded rivers.

(b) Silver Carp (*Hypopthalmicthys molitrix*)

Class	:	Teleostomi
Order	:	Physostomi
Family	:	Cyprinidae
Genus	:	*Hypopthalmicthys*
Species	:	*molitrix*

Silver carp naturally occurs in the river systems of central and south China and in Amur river of USSR, and now it is cultured in many countries including India.

This was brought to India in 1959 from Japan for culturing at CIFRI, Cuttack and Orissa. In due course, it was bred successfully by hypophysation and the fries were distributed from Cuttack to various States of India.

The diagnostic characters of silver carp are the oblong and slightly compressed body, head pointed, bluntly rounded snout, slightly protruding lower jaw, small scales, abdomen keeled from the throat to the ventral, origin of dorsal fins behind ventral and dark fins.

In case of this fish, fry and adult sustain on *Flagellata. Dinoflagellata*, *Myxophycea*, *Protozoa* and *Rotifera*, supplemented with decayed macro-vegetations and detritus.

When kept in ponds, it thrives well on artificial feeds, such as bone meal, rice bran and flour.

The fish attains full maturity by the time they are two years old.

This is an esteemed food fish and shows encouraging production when stocked with other indigenous carps.

(c) Common Carp (*Cyprinus carpio*) (Figure 6)

Class : Teleostomi

Order : Physostomi

Family : Cyprinidae

Genus : *Cyprinus*

Species : *carpio*

This is a world wide most extensively cultivated species.

Although exotic of India, it is already being cultivated in the country singly as well as along with Indian major carps.

The fish is voraciously omnivorous, efficiently converting the food ingested into flesh, grows very fast and prone to artificial feeds.

It is non predatory, when cultured along with major carps, it competes for food mostly with mrigal and kalbasu.

The strain of common carp was obtained at Cuttack from Bangkok in 1957, is found to procleate prolifically in the plains of India and has been distributed to different States of the Country.

The common carp has three recognized varieties, differentiable according to the pattern of scaling.

(*i*) The Scale Carp (*Cyprinus carpio var.* communis)

Class : Teleostomi

Order : Physostomi

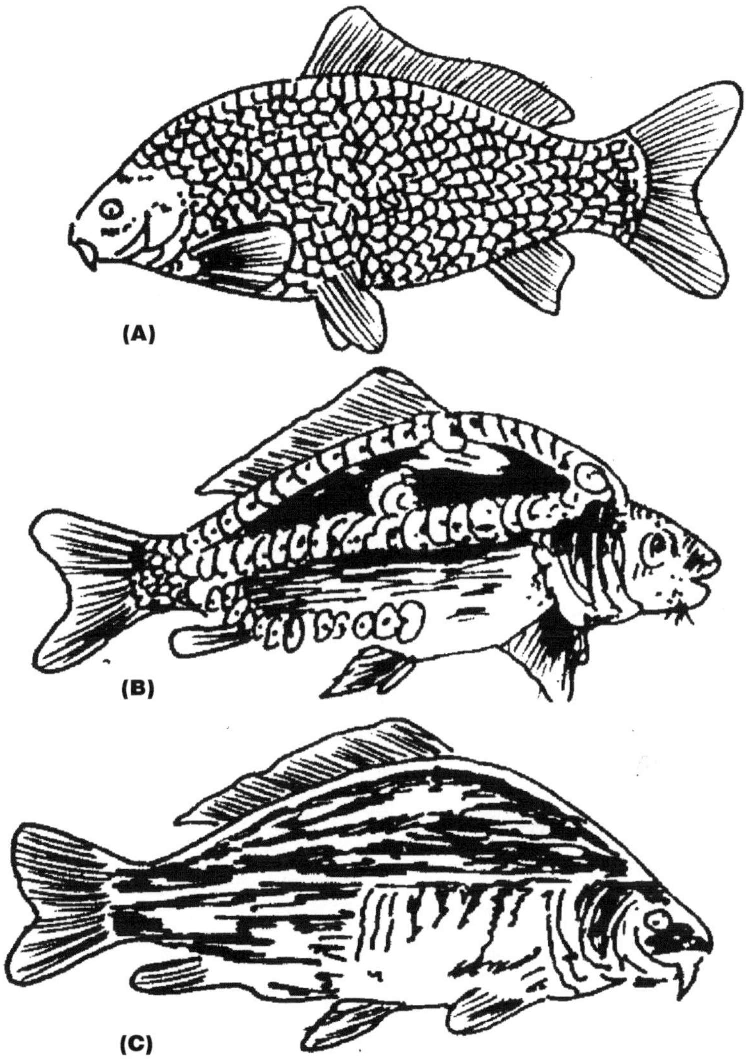

Figure 6: Common Carp (*Cyprinus carpio*)
(A) Scale Carp (*Cyprinus carpio var.* Communis)
(B) Mirror Carp (*Cyprinus carpio* var. specularis)
(C) Leather Carp (*Cyprinus carpio* var. nudus)

Family : Cyprinidae

Genus : *Cyprinus*

Species : *carpio*

Variety : communis

It has its body fully covered by regularly arranged rows of scales.

This is the exotic form, now extensively cultivated in our country in the various States including Haryana.

(*ii*) The Mirror Carp (*Cyprinus carpio* var. specularis)

Class : Teleostomi

Order : Physostomi

Family : Cyprinidae

Genus : *Cyprinus*

Species : *carpio*

Variety : specularis

It has its body covered unevenly with a few large and bright scales.

A large area of the body is, however, naked.

This species is extremely variable and is also cultivated in the State of Haryana.

(*iii*) The Leather Carp (*Cyprinus carpio* var. nudus)

Class : Teleostomi

Order : Physostomi

Family : Cyprinidae

Genus : *Cyprinus*

Species : *carpio*

Variety : nudus

It has its body completely devoid of scales except for a single row of somewhat degenerated scales along the base of the dorsal fin and sometimes extending as far as the tail or even from head to tail.

Sometimes tiny scales are also present at the bases of other fins.

Chapter 4

Examination of Gonads and Eggs of Culturable Fishes

There are several criteria for the determination of maturity stages in fishes, which vary from one group of species to another. However, in common carps the state of maturity of fish can be determined in field examination by exposing the gonads in the body cavity. Maturity stages in these fishes can be examined as follows.

Virgin: Very small sexual organs located near the vertebral column, Testis and ovary transparent, colourless to gray. Eggs invisible to naked eye.

Maturity Virgin and Recovering Spent: Testis and ovary translucent, gray-red. Length half, or slightly more than the length of ventral cavity. Individual egg can be seen with magnifying glass.

Developing: Testis and ovaries opaque, with blood capillaries. Occupy about half of ventral cavity. Eggs visible to the eye as whitish granules.

Developed: Testis reddish white. No milt drops appear under pressure. Ovary orange-reddish. Eggs clearly discernible; opaque. Testis and ovary occupy about two thirds of central cavity.

Gravid: Sexual organs filling ventral cavity. Testis white, drops of milt fall with pressure. Eggs completely round, some already translucent and ripe.

Spawning: Roe and milt run with slight pressure. Most eggs translucent with few opaque eggs left in ovary.

Spent: Not fully empty. No opaque eggs left in ovary.

Resting: Testis and ovary empty, red. A few eggs in the state of reabsorption.

The gonads and egg counts are further examined as follows:

The gonads should be removed and weighed in fresh condition, because their weight may change considerably when preserved. The counting of eggs can however be done on preserved material. When counting eggs in the field, measure the total displacement volume of the eggs from the ovary in a graduated cylinder. Remove the eggs from the water and count a portion of them.

Measure the displacement volume of the counted eggs and compute the number as follows:

$$\text{Number of Eggs} = \frac{\text{Total Volume of eggs} \times \text{Number of eggs Counted}}{\text{Volume of the counted eggs}}$$

The diameter of fresh eggs may be measured by placing 10–50 eggs in a row, measuring the total length of the row, and dividing by the number in the row.

In life history studies and in fish management, the knowledge of the numbers of eggs produced by each species is essential. The number of eggs must be known if survival is to be estimated and the number produced is significant in fish culture in determining the number of fish required for brood stock. Egg count can be made in two ways. First, at the time of spawning when the eggs are fully developed and ready to be laid. The females can be captured and, with pressure applied to the abdomen, the eggs can be forced from the body into a pan or jar for preservation and enumeration. If it is not necessary to save the female for some other purpose, the female can be killed and cut open and the eggs remaining after the stripping operation can be added to those stripped from her. If the female cannot be killed, then the number of eggs must be considered to be approximate until such time as it is possible to determine the average quantity of eggs remaining after normal stripping operation.

The second method is to remove the entire ovary from the fish prior to full ripening. The closer to the time of full ripening it is possible to obtain the ovaries; the better it is for counting.

The most accurate enumeration of eggs is the actual count, but this can be very tedious and time-consuming. When actual counts become impractical (as in species with very large numbers of eggs) approximate number can be obtained by one of the following methods.

Volumetric Method

Under this method there are two ways. First count the eggs in a sample of known volume (v) taken from the unknown lot. Such a sample may be measured by water displacement in a finely graduated cylinder and 10 or 20 ml should be adequate. Then measure the total volume (v) of all eggs in the unknown lot. The total number can be estimated as follows:

$$X : n = V : v$$

where,

X = Total number in the unknown lot

n = Number in sample

V = Total volume of all the eggs

v = Volume in sample

The second method is to count out 100 eggs and measure the volume by water displacement. If the eggs are small, count out 200 or even 300 eggs and obtain the volume. Then the entire sample is measured by displacement, and the total number of eggs would be calculated by the formula given above.

Gravimetric Method

First determine the total weight of the whole lot of eggs and calculate on the basis of the sample counting by weighing a definite portion the sample, such as 1 or 2 grams, and then count the eggs in the sample. The second way is to weigh a known number of eggs, and then the total lot, and calculate the total number.

In the gravimetric system it is important to remove the excess moisture. This may be done by placing the eggs on blotting paper for a given length of time, or by spreading the sample in a large pan and allowing the eggs to be exposed to the air for a definite period of time, or in some cases the samples may be placed in a drying oven for a time.

Von Bayer Method

In 1910, Von Bayer prepared a table for determining the number of eggs per liquid quart based on the diameter of the eggs. A metal through of known length, usually 150 mm is used. The groove in the trough is filled with a single row of eggs, and the number of eggs it takes are counted. The diameter of the eggs is determined by dividing 150 mm by the number of eggs. By referring to his chart the number of eggs per quart is found. These values may be changed into number per ml by dividing the number per quart by 946.4 (the number of ml per quart). By determining the number of ml of eggs in the sample by displacement, and multiplying the number found by the number of eggs per ml, the total number of eggs found. Von Bayer's chart is reprinted in Lagler's Freshwater fisher biology.

Egg Counts in Preserved Ovaries

Measure the total volume of the ovaries by water displacement. Then obtain volumes of each of three sections, well-spaced, from the ovary. Separate out and count the mature eggs from each ovary section, separating them from the immature eggs and tissues. Take the volume of immature eggs and tissues and subtract from the volume of the section to get the volume of the mature eggs. Take the average results of three (or sometimes more) count and compute the number of fully developed eggs per ml and then for the complete ovary.

The egg count can also be made using the gravimetric method by weighing three samples from various parts of the ovary, counting the eggs as described above, and calculating the total numbers on a basis of weight instead of volume.

Chapter 5

Identification of Fish Seed of Culturable Species of Fish from Pond (Spawn, Fry, Fingerlings)

Fish hatchlings measuring up to 8 mm in length are known as spawn. Fish seed above 8 mm and up to 40 mm in length are known as fry. Fish seed above 40 mm and upto 150 mm in length are known as fingerlings.

A typical fish pond consists of three food niches: surface, column and bottom. Different varieties of fishes feed in different niches of the pond. For example, Catla is surface feeder. Rohu is column feeder and Mrigal is a bottom feeder. While stocking the ponds with fry and fingerlings different groups of fish with different feeding habits are stocked so that food available in all the three niches is properly utilized for maximum growth of fish seed.

The following are identification characteristics of different species of spawn, dry and fingerlings Carp spawn show jerky forward movements, whereas, the spawn of catfishes exhibit wriggling movements.

Carp spawn, when it is 5.0 to 5.5 mm long is differentiated into desirable and undesirable varieties. In the desirable carp spawn, the yolk sac is wide anteriorly, the embryonic fin fold is continuous, the eyes pigmented and the mouth is formed. In the undesirable carp spawn, the yolk sac is disappearing, rudimentary or absent. The embryonic fin-fold is discontinuous, there are rudiments of dorsal fin with or without fin rays, the mouth is formed and these are pigmented.

Carp fishes are divided, broadly into (*i*) major carps and (*ii*) minor carps. Major carps grow faster and are bigger in size as compared to minor carps. At the fry stage, the carps can be easily differentiated into major and minor.

Major Carps

The number of undivided dorsal fin rays is more than eleven (*e.g. Catla catla, Labeo rohita, L.finbriatus, L.calbasu, L.gonius, Cirrhinus mrigala*).

The Major Carps Consisted of Catla, Rohu and Mrigal

Individual characteristics of major carps fry of different species are as under:

Catla catla

Large head, Dorsal profile convex and ventral profile concave. No distinct spot on the caudal fin, or at the caudal peduncle, Margins of the dorsal and caudal fins darker. First ray of the dorsal fin is black. Opercular region brightly reddish. No barbels. Lips thick but not fringed.

Labeo rohita

A dark diffuse transverse dark band present at the caudal peduncle. A pair of whitish or grayish maxillary barbels present, lips fringed.

Labeo calbasu

At the earlier stages, the fry are darker in colour. A semilunar yellow band appears on the base and gradually the body gets the appearance of having alternate yellow and black bands. The basal portions of the fins are black which gradually spread over the fins. Two pairs of thick black barbels (maxillary and rostral) present. Lips fringed, mouth narrow.

Labeo gonius

Body marked with a number of very faint longitudinal white stripes. An indistinct spot present at the caudal preduncle. A pair of maxillary barbels present. Lips fringed, mouth narrow.

Labeo fimbriatus

Head small and mouth terminal. A conspicuous dark caudal spot present, small head with thick fringed lips. A pairs of barbels present. Lips fringed, mouth narrow.

Cirrhinus mrigala

Small head and slender body. A more or less triangular dark spot at the caudal peduncle. No barbels visible. Lips thin, not fringed.Tip of the lower lobe of the caudal fin reddish.

Minor Carps

The number of undivided dorsal fin rays are eleven or less than eleven (*e.g. Labeo bata, Cirrhinus reba and Puntius surana)*

Labeo bata

A bright dark caudal spots more or less semi-circular in shape present at the caudal peduncle. This spot gradually becomes more or less circular in outline. No barbels visible. Lips not fringed.

Cirrhinus reba

Slender body. Presence of three distinct dark spot: two on the caudal fin and one at the base of the caudal peduncle. Pigmented spots on the lateral line, which gradually take the shape of a dark lateral band. Mouth terminal, no barbels, not fringed.

Puntius surana

Body deep and moderated compressed. A small black vertical band extending from the point of insertion of the dorsal fin. An oval dark spot present at the caudal peduncle. Lips thin and not fringed. Maxillary barbels present. The first ray of the dorsal fin blackish.

Materials Required

Spawn, fry and fingerlings of different species.

Formalin

Glass bottle with a cork

Petri dishes

Beakers

Enamel trays

Plastic buckets

Mug

Freshwater

Pipette

Fine brush

Forceps

Hand lens

Hand and drag nets

Pencil and eraser

A few drawing sheets

Procedure

Take a sample of fish seeds in a beaker and observe the movements. (Generally fish seeds showing wriggling movements are catfish seed; and those showing jerk, forward movements are carp seed).

Take a Petri dish with formalin and place a sample of fish seed in it.

Measure the length and decide whether the sample consists of spawn, fry or fingerlings. If the sample consists of carp seed, count the dorsal fin rays and determine whether they belong to major carps or minor carps.

Study of Individual Characteristics

Make diagram of different species with a pencil on the drawing sheets and note down individual characteristics. Name the individual species by drawing inferences from individual characteristics.

Precautions

Handle the spawn, fry and fingerlings carefully so that they are not injured.

Stock only healthy fry and fingerlings into the ponds.

Measure length of spawn, fry and fingerlings correctly for differentiation into the respective groups for stocking purpose.

Chapter 6

Stocking of Spawn and Fry in Nursery Ponds

Spawn is a tiny fish (5-6 mm in length) in which the yolk sac is absent and the digestive system is developed and it performs horizontal, free movements in the water. This spawn up to 9 mm size is stocked in the nursery ponds and after it has attained the size of 30-40 mm, it is called fish fry. The fry can be stocked in the rearing ponds to grow up to fingerling size (40-50 mm). Nursery ponds are for the stocking of fry.

Nursery ponds are the small water bodies in which proper care of feeding and other necessary steps can be undertaken easily. The merits of stocking spawn and fry in nursery and rearing ponds respectively are as under:

Spawn or fry can get the feed easily;

More seed can grow in a given area under proper care;

Spawn and fry grow faster;

It is easy to remove the predators and other harmful organisms from nurseries;

Survival percentage from spawn to fry and fry to fingerlings is high.

It is easy to harvest the seed.

Assessment of seed can be done in nurseries by simple netting.

Generally, the spawn is stocked at the rate of 50-60 lakhs/hectare as a conventional rate of stocking but under modern nursery management with the help of aeration devices and recirculation of the water systems and proper feedings, it has become possible to stock 1.5 to 2 crores of spawn/hectare with a high percentage of survival. In the case of fry, the general stock rate is 2.5-2.0 lakhs per hectare but under modern technologies, it is possible to increase the stocking density.

Counting of Spawn/Fry

Spawn are measured by measuring cups which are perforated, cup-like structure of different size (5, 10, 20, 50 ml), made of metal. Two or three samples of spawn are taken and the number of spawn in each sample is counted. Then the mean of the samples is derived and that is the number of spawn in the cup. In the case of fry, the same procedure is followed, but the size of the cup is bigger. The general stocking rate of spawn of carps is 60 to 70 lakh/ha, and 6-8 thousand/ha, in case of fingerlings.

Material Required

Measuring cups for spawn/fry.

Tub, buckets, mugs, for transfer of spawn and fry.

Watch glasses.

Muslin cloth.

Procedure

Calculate the stocking density as per the nursery or rearing pond area. Take out the spawn from the hatchery or other breeding grounds in a hapa and keep in a tub filled with water.

Arrange a number of buckets with water (5-6 litres) according to the availability of spawn.

Measure the spawn from the measuring cups and keep in the buckets (2 lakhs in each).

Take the buckets to the nursery site.

Enter the nursery pond up to knee level and dip half of the bucket in the water and then wait for 5 to 10 minutes.

Keep the bucket in the water in such a way that the spawn/fry can get in the pond.

Wash the bucket properly in the nursery pond.

Observations

The student should observe and record the following:

Mortality of spawn or fry.

Movement of spawn/fry along the periphery of the nursery ponds.

The pupil should maintain a diary.

Precautions

Take 3 to 4 samples of spawn and count to find out the average number of spawn/fry per measuring cup.

Handle the spawn or fry carefully to avoid mortality.

Stock the nursery ponds with spawn preferably in cool water.

Do not release the spawn and fry at a time in one pond but release them separately into the nurseries/rearing ponds.

If an oil layer is present on the surface after the application of oil emulsion in the nursery pond, remove it by splashing or disturbing the water or with the help of cotton cloth.

Do no release all the spawn/fry at one place in the nurseries/rearing pond.

Chapter 7

Natural and Supplementary Feeds of Cultivable Fishes and Estimation of Feed Conversion Ratio (FCR) and Food Conversion Efficiency (FER)

Catla (*Catla catla*)

Catla is a surface and column feeder and its upturned mouth and large gill rakers are adapted to feed on the numerous organisms floating in water. Submerged aquatic weeds are not of any value as food for catla, and in ponds with overgrowth of submerged weeds; its growth is generally unsatisfactory. Young ones, from the time they begin feeding until they reach a length about 15-20 mm, feed almost exclusively on water fleas, *Cyclopes* and *Daphnia* and other animalcules in the water. After this stage, they are probably capable of making use of microscopic plants, floating in the water, though at all stages, water-fleas and other animalcules in the water, constitute an important item of food.

Rohu (*Labeo rohita*)

Rohu is a column-bottom feeder and rather an inferior fringed-lipped mouth is well adapted to browsing on shallow pond bottom. Considerable quantities of bottom sand, vegetable debris or decaying leaves of aquatic plants, planktonic algae etc., constitute the stomach contents. Water fleas and other animalcules are rarely found among the food items consumed by this fish.

Mrigal (*Cirrhina mrigala*)

Mrigal appears to be a bottom feeder, thought her young ones, as soon as they commence feeding have the same feeds as those of catla and rohu. The fingerlings and adults take in relatively large quantities of decaying organic and vegetable debris, phytoplankton organisms, and sand and mud as compared to rohu. The proportion of animal matter in diet is generally poor. The thin terminal lips are adapted to picking up things from the mud and from the invariable presence of appreciable quantities of sand and mud in the gut, it is presumed that algae in the fresh or decaying conditions are also picked up from the bottom.

Kalbasu (*Labeo calbasu*)

Kalbasu feeds on snails and worms at the bottom of the pond, in addition to usual algae and other elements on which rohu and mrigal feed.

Seenghara (*Mystus seenghala*)

This is a predatory and piscivorous fish, mainly feeds on small and medium sized other fishes. At the younger stage, this fish feeds on water fleas, insects, other fish fry and fingerlings of carps.

Live Fishes

Ophiocephalus punctatus

This is a predatory fish very much common in our freshwater bodies. It feeds on small fishes, aquatic insects, micro crustaceans, shrimps; the young ones feed exclusively on planktonic crustaceans and insects.

Ophiocephalus striatus

It is largest among the murrel fish (4′). It is highly predacious and even cannibalistic, feeding on small and medium sized fishes. The young ones feed on insect larvae and fish fry. It is very dangerous to culture them along with the other fishes.

Ophiocephalus marulius

The food and feeding habits of *O. marulius* and *O. gachua* are almost similar to those of the above two fishes of family Ophiocephalidae.

Various Types of Common Supplementary Fish Feeds

1. Mustard oil cake (MC)
2. Groundnut oil cake (GNC)
3. Soybean flour
3. Soybean cake
4. Rice-bran
5. De-oiled rice bran
6. Maize flour
7. Wheat flour
8. Barley flour
9. Kitchen-vegetable wastes
10. Butcher house animal offal (Non-conventional feedstuff)
11. Pelleted dry feed
12. Pelleted wet feed
13. Fishdry fish feed
14. Prawn-dry fish feed
15. Aquatic weeds–feed for grass carp

Food Intake Status in Culture Fishes

Normally, all the feed should be consumed in 7-8 hours. If all the food is not consumed in this time, the fish have an abnormal ingestion status and the quantity of feed should be reduced. The most suitable size of fish feed at different stages are 0.07 mm (for spawn), 0.08 mm (for fry) and 1.2 to 2.4 mm pellet size for fingerling and others. Fixed quality, quantity, time and position of feed are important in a proper feeding schedule. The particle size should be 100-200 micron (μm) for larval fishes.

Fixed Quality

Feed should be fresh and palatable with a high nutritive value. Spoiled foods should be discarded to prevent diseases.

Fixed Quantity

Fish should be provided with a fixed amount of food every day. Uneven feeding will cause poor digestion, bad absorption and slow growth rate.

Fixed Time

Feeding is usually done before 8 to 10 am in the morning and between 4 to 6 pm in the evening every day. To raise the food utilization rate, fish should be fed when the dissolved oxygen content is high.

Fixed Position

Feeds such as pellets and fine feeds should be supplied at a fixed position (*e.g.* on a fixed feeding platform/basket/polythene bag). This allows easier examination of the fish. Floating feeds may be supplied in a floating framework.

Feeding Rate

Carps are herbivorous but having different habits. The rate of feeding depends upon the biomass of the cultured species which should be adjusted by having observations on their daily feed intake. Overfeeding results in feed wastage as well as deterioration in water quality and food conversion ratio (FCR). Based on the studies, the feeding rate of fry, fingerling and adult are summarized here (Tables 7.1–7.3)

Table 7.1: Feeding Rate of Cultured Carp Species (Average)

Stage	*Spawn*	*Fry*	*Fingerling*	*Adult*	
				1 yr.	*II yr.*
Ist five days	Equal to double the body weight	5–10%	2–5%	5–6%	2–3%
6th to12 days	Double to thrice of the body weight				
13th day	No feeding				
14th day	Harvesting (for stocking) (Feeding can be adjusted with the weight of fish)				

Size: Spawn, 6–7 mm; fry, 14–25 mm; Fingerling, 40–100 mm; Adult, 300–400 mm.

Grass carp consumes soft leaves of aquatic plants equal/double the weight of the body of fish.

Table 7.2: Feeding Schedule for Grass Carp Yearling

Fish Size (cm)	*Feed*	*Feeding Duration (Daily)*	*Daily Feeding Rate (kg/10,000 fish)*
3–7	*Wolffia arrhiza*	15–20% (from June)	20–40%
7–8	*Hydrila verticillata* and *Lemna minor*	15–20% (from July)	75–100%
8–10	*Spirodella polyrrhiza*	15–20% (from July)	100–150%
10–13	*Vallisneria spiralis*	40–50% (from Aug.)	150–200%
13	*Vallisneria spiralis* and commercial feed	80–90% (from Sept.)	75–150%

Evaluation of Nutritional Value of Fish Feeds

In feeding, it is important to know the nutritional requirements of the fish and also evaluation of the nutritional value of the feeds and their ingredients. Besides the chemical analysis, nutritional value of fish feeds must be evaluated with other criteria such as digestibility, utilization rate and food conversion ratio via feeding trials.

Digestibility: The degree to which food stuff is broken down into particles or forms suitable for absorption following ingestion.

$$\text{Digestibility (\%)} \quad \frac{\text{Nutrient intake} - \text{Nutrient remaining in faeces}}{\text{Nutrient intake}} = \times 100$$

Food utilization ratio – refers mainly to the utilization rate of the crude protein in the diet.

$$\text{Utilization rate of Crude protein (\%)} = \frac{\text{Protein increase in fish body}}{\text{Protein in consumed food}} \times 100$$

Food Conversion Ratio (FCR)

In fish production, calculating the food conversion ratio is a common method to appraise nutritional value. In general, it is lower for the juveniles than for the adult fish.

Table 7.3: Feeding Methods

Type of Feeding	*Type of Feed*	*Usage*	*Advantages*	*Disadvantage*
Hand-feeding	All types of supplementary and intensive feeds	Used in Semi-intensive culture	No capital expenditure on feeders	Can be expensive in man hour's at large farms. Poor FCR values and high waste loading.
Demand feeder	Pelleted dry feeds	Used in Semi-intensive culture	Cheaper than automatic feeders and less labour intensive than hand feeding. Rates can be adjusted to suit conditions.	Care must be taken to avoid over-feeding small fish (10 g can't use most designs). Poor feed distribution.
Automatic feeder	Pelleted dry feeds	Used in intensive culture of salmonids	Farmer can set feeding rate and duration	Noisy
(a) Air operated		Not commonly used in cages	Relatively good feed distribution, less labour-intensive.	No feedback from fish to farmer hence, risk of over-feeding
(b) Electrically operated	Pelleted dry feed	Widely used in intensive fish culture	Farmer can set feeding rate and duration. Less labour intensive than hand feeding	No feedback from fish to farmer, hence risk of over feeding
(c) Water operated	Pelleted dry or wet feeds	Used with wet feed for salmon culture	Computerized system responsive to environmental changes, improved FCR values and reduced solid loading. Computerized comprehensive records.	May be problems with vitamins.

$$\text{Food conversion ratio} = \frac{\text{Food consumed (Dry weight)}}{\text{Fish body weight gain (Wet weight)}}$$

$$= \frac{\text{Food applied}}{\text{Total weight at harvest} - \text{Total weight at stocking}}$$

The food conversion ratio is affected by many factors, particularly management. There are mainly five factors that affect the food conversion ratio

Food preparation

Feeding regime

Water quality

Nutritional element in the food

Species and age

Food Conversion Efficiency (FER)

It is the inverse of the food conversion ratio

$$\text{Food conversion efficiency} = \frac{\text{Total fish body weight gained}}{\text{Total weight of feed intake}} \times 100$$

Growth

Growth is the increase in body mass per unit of time. It depends on the proper nutrition and management. It should be measured before stocking the fish and at the time of final harvesting, followed by weekly/fortnightly intervals to regulate the feeding and healthy growth and better utilization of feed.

The following formulae are used in growth studies of fish.

$$\text{Fish growth rate} = \frac{\text{Total gain of fish (g)}}{\text{No. of fingerlings and their culturing days}}$$

$$\text{Daily weight gain} = \frac{\text{Total weight gain of fish during the rearing period}}{\text{Total quantity of the fish and their culturing days}}$$

Precautions

Container should be cleaned before and after feeding and washed in lime water.

Avoid excess feeding and supply feed regularly in required quantity.

In ponds, infested with algae, stop feeding and remove the algae.

Chapter 8

Feed and Nutritional Requirement of Pond Fishes

Every organism needs energy for all its activity. This energy is obtained through food. The process by which an organism obtained nutrient is called as the nutrition.

In aquaculture system for fish production, especially for Indian major carps, *Catla catla, Labeo rohita, Cirrhinus mrigala* and exotic carps, *Ctenopharyngodon idella, Hypophthalmichthys molitrix* and *Cyprinus carpio* scientifically formulated and optimally processed, nutritionally balanced, eco-friendly and economically viable artificial feeds are essential which can be used as supplementary or complementary feed in fish farms.

Nutrition

It is defined as "Series of processes by which an organism takes in and assimilates feed for promoting growth and replacing dead or injured tissues."

What are the Nutrients?

The chemical substances found in feed materials are necessary for the maintenance, reproduction and health of animals. The chief classes of nutrients include 25 carbohydrates; 20 amino acids; 15 fatty acids; 20 vitamins; probably 10 essential minerals and water.

Feed Constituents

For proper growth and good health, fish requires balanced diet, which is made up of organic and inorganic ingredients, referred to as food constituents. According to their functions in the body they are:

Body builders

Energy producer

Regulators

Proteins

Carbohydrates

Vitamins

Minerals

Fats

Salts

Water

Effect of Mineral Deficiency on Fish

Mn Reduced growth, loss of equilibrium, short body, dwarfism high mortality, poor hatchability of eggs, abnormal tail growth.

Cu Reduced growth, anemia, muscular dystrophy

I Thyroid hyperplasia (Goiter)

Cr Required for carbohydrate and lipid metabolism. Its function is related to insulin.

F Enhance fluoride accumulation in vertebrae

Mo Improves growth and survival

S Supplement growth and provide amino acid requirements

B Boron improves growth and survival.

Desirable Characteristics of a Diet

The diet should be properly balanced.

The food must be palatable.

Variety of feed in the diet.

The feeds composing the ration should be good.

The ration should contain enough of mineral matter.

Table 8.1: Vitamin Requirement and Deficiency Signs

Vitamins	*Requirements mg/kg/dry*	*Deficiency Signs*
Thiamin (B_1)	2–3	Fin congestion, nervousness and fading of body colour
Riboflavin (B_2)	7–10	Skin and fin haemorrhage, mortality
Pyridoxine (B_6)	5–10	Nervous disorders
Pantothenate (Pantothenic acid)	30–40	Poor growth, anaemia, skin haemorrhage, exopthalmina, clubbed gills.
Niacin (Nicotinic acid)	30–50	Skin haemorrhage, mortality
Folic acid	15	Poor growth, lethargy, dark colouration
Cyanocobalamin (B_{12})	0.09	Poor appetite, low haemoglobin
Inositol	200–300	Skin lesions, dark colour
Choline	500–600	Fatty liver, poor food conversion
Ascorbic acid (Vitamin–C)	30–50	Scoliosis and lordosis and impaired collagen formation
Retinol (Vitamin–A)	1000–2000 IU/kg	Faded colour, exopthalmia, fin and skin haemorrhage, poor vision, and abnormal bone formation.
Vitamin D (D_3)	500 IU/kg	Reduced weight gain, lethargy, poor appetite, abnormal bone formation.
Cholecalciferol –tocopherol	80–100	Muscular dystrophy
(Vitamin–E)	IU/kg	Mortality, poor lordosis
Menadione (vitamin K)	40 IU/kg	Impairment of the blood–clotting mechanism
Biotin (vitamin–H)	1–15	Poor growth, loss of appetite, skin lesions, poor growth, de–pigmentation, maximum growth rate, increased production per unit area per unit time, increased reproductive efficiency and minimum mortality
Basic composition of feed	–	The major components of feed stuff are moisture, lipid, protein, fibre, carbohydrate, minerals and vitamins
Practical feed formulation	–	Formulation of practical fish diets for intensive aquaculture is usually based upon a knowledge of the nutritional requirements for the fish species under cultivation, in particular the optimum protein and energy levels and the inclusion of essential amino and fatty acids, together with vitamins and minerals.

The ration should be fairly laxative.

Avoid sudden changes in the diet.

Maintain regularity in feeding.

The feed must be properly prepared.

Economy in labour and cost of feeding.

Feed Concentrates

All feeds supplying primary nutrients (protein, carbohydrates and fat) are classed as concentrates. Their crude fiber content does not exceed 18 per cent. Concentrates are normally of either plant or animal origin as shown in Table 8.2.

Table 8.2: Major Sources of Primary Nutrients

Plant Sources	*Animal Sources*
Maize	Fish meal
Barley	Blood meal
Oat	Meat meal
By-products of various grains	Prawn meal
Wheat bran	Silkworm pupae
Rice bran	Poultry waste
Rice polish	Acetes
Oil cakes	Bone meal
Groundnut cake	
Mustard cake	
Linseed cake	
Tile cake (mustard)	
Soybean cake	

Feed Supplements

The feed stuffs which are used to improve the value of basal feeds are known as feed supplements. They can be used in large quantities such as protein supplements or in extremely small quantities, such as trace elements/minerals, calcium carbonate, sodium chloride, cobalt chloride, potassium iodide, ferrous sulphate, magnesium sulphate, manganese chloride, phosphoric acid, potassium chloride, ammonium bicarbonate, zinc chloride.

Selection of Feed Ingredients

A large number of viable feed ingredients can be profitably utilized for formulating fish diets. However, the selection of ingredients should be done very carefully. It should be proteinaceous, locally available, cheaper ingredients of either animal or plant origin having stability during processing, storage and feeding. The proximate composition of certain commonly available ingredient, are given in Table 8.3.

Table 8.3: Proximate Composition of Common Fish Feed Ingredients

Ingredients	*As per cent Dry Matter*			
	Protein	*Fat*	*Fibre*	*Ash*
Rice bran (de oiled)	15–17	1–2	14–18	16–20
Groundnut cake	41–43	3–8	6–8.5	4–8
Soybean cake	42–48	2–7	6–8	5–7
Fish meal	52–60	5–12	1–4	22–38
Prawn meal	58–65	4–7	4.7	21–26
Blood meal	79–88	1–2	–	4–6
Brewer's yeast	4.2	1.4	3.0	6.8
Wheat bran	12–14	6–8	10–14	6.8
Silkworm pupae	48–53	26–30	6–8	7–11
Poultry waste	52–59	16–35	–	9–18

Feed Additives

An additive is a substance that is added to a feed, usually in small quantities, for the purpose of fortifying it with certain nutrients, stimulants or medicines other than a direct source of nutrients.

A number of feed additives are in use to protect the labile nutrients, improve the quality of the feed, maximum utilization of feed and availability for longer duration to the animal. These additives may be binders, chemo-attractants, antimicrobial agents, antioxidants, growth promoters, odours, flavours etc. The selection and inclusion levels of these additives are dependent on their effectiveness and cost. The recommended levels of some tested additives are given in Table 8.4.

Table 8.4: Some Important Additives Used in Fish Feed

Binders (Recommended quantity)	*Chemo-attractants*	*Antimicrobial Attractants*	*Antioxidants*	*Pigments*
Guar gum (1–2%)	L–alanine	A. Preservatives (useful in controlling fungi)	Ethoxyquin (0.015% or 150 mg/kg)	Carotenoids Zeaxanthin Cryptoxanthin (0.5–1%) Astaxanthin Canthaxanthin
Gum-acacia (2–5%)	Glycine Hydro-chloride	Sodium and calcium propionate	Butylated Hydrox-Anisole (0.02% of the fat)	Astaxanthin Canthaxanthin
Alginates (2–5%)	Trimethyl Ammonium Hydro-chloride (TMAH)	Propionic acid		
Wheat gluten (10–12%)				
Carrageenan and agar (2–5%)		Sodium benzoate (0.1– 0.25%)		
Starch from wheat flour and tapioca (2–5%)		B. Antibiotics useful in controlling disease (dose: 5.5– 8.25 mg/100g of fish/day for 10 days) Oxolinic acid, Oxytetracycline, or Teramycin		
Casein Chitosan Cornstarch Potato starch Hemicellulose (2–5%) Bentonites (2–3%) Molasses (2–3%) Whey (1–3%)				

Table 8.5: Different Types of Feed Depending on Moisture Contents

Dry feed	*Semi-moist feed*	*Moist/wet feed*
It contains 7–9% moisture	It contains 7–10% moisture	It contains 7–13% moisture
It can be prepared either by using dry ingredients alone or a mixture of wet and dry ingredients by processing	It is made from a mixture of high moisture ingredients and dry pulverized ingredients	It is prepared from high moisture ingredients such as trash fish, fishery waste, slaughter-house waste, etc.
It can be made as meals, pellets, granules, flakes, etc.	More palatable to most species	Advantages and disadvantages are same as the moist feed.
It has poor water stability and cause water pollution due to breaking up and dissolving of feed in water.	It is more prone to introduction of pathogens, damages of certain heat labile vitamins and lipids and flavour. Propagation of fungi and bacteria in such feed.	

Other Types of Feeds

Natural food–Produced by natural water bodies.

Artificial feed

A diet manufactured from grains, fish meal, meat and meat by-products or purified ingredients, as added to a natural diet of living organisms or detritus found in the aquatic environment also called prepared diet.

Based on the requirement of fish, feed are of the following types:

1. Balanced Feed

A feed having all the required nutrients in the proper amount and proportion to meet the needs of the animal body is called a balanced feed.

2. Complete feed

An aquaculture diet that satisfies all the nutritional requirements of the target species or a nutritionally adequate feed for animals which is compounded by a specific formula, to be fed as the sole rations and is capable of maintaining life and or promoting production without any additional substances being consumed except water is called a complete feed. It includes both roughage (whole plant materials of high fiber content which form the bulk of the rations fed to animals) and concentrate components.

3. Compounded Feed

A compounded feed is composed of a number of feed ingredients. It could be a concentrate mixture or balanced feed.

4. Purified Diet

A prepared diet composed of individual amino acids, fatty acids, simple sugars, vitamins and minerals.

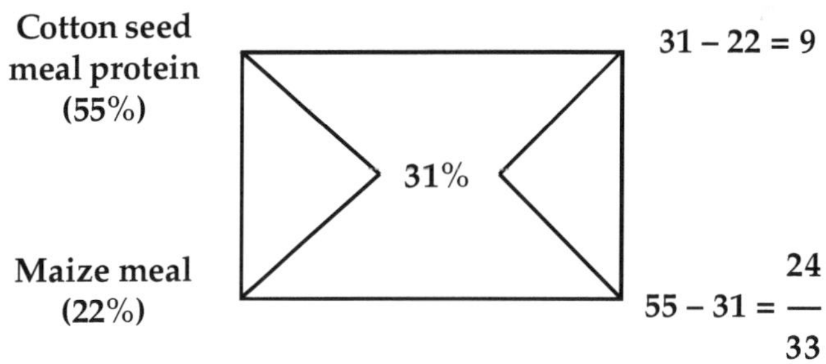

5. Supplementary Food

A prepared diet formulated to provide protein and other nutrients in excess of those obtained from natural food organisms in the environment.

6. Pelleted and Mesh Feeds

Feeds are either produced in powdered/mesh or pelleted form.

Advantages of Pelleted Feeds

Pelleting prevents the ingredients from segregating.

It does not allow the animal (fish) to sort out certain ingredients and reject others.

It reduces wastages of feed and loss of nutrients during storage.

Pellets are less subject to infestation of insects and moulds.
It reduces the possibility of adulteration of feeds with undesirable substances after pelleting.

It is easy to handle.

Pelleting kills unhygienic bacteria due to exposure to high temperature during processing.

It increases the palatability and digestibility of feed.

It reduces to a certain extent the microbial degradation of protein and preserves vitamin-potency.

The Quantities are Calculated as Under

The ingredient protein percentage is written on left side of the square and desired percentage at the middle. Subtracting the middle value diagonally from higher to lower irrespective of minus sign, calculate, the differences are written on the right side.

$$\text{Cotton seed meal} = \frac{9}{9+24} \quad L \times 100$$

$$= \frac{9}{33} \times 100 = 27.27\%$$

To prepare 100 kg feed we need 27.27 kg cotton seed meal and 72.72 kg maize germ meal to get 31 per cent protein in prepared feed. Similarly, essential fatty acids and essential amino acids are also matched to obtain the desirable level.

Chapter 9

Manuring and Fertilization of Nursery Ponds

Manuring is an important step in the management of nursery ponds for better growth in the rearing of carp spawn. Organic manures like cow dung and various oilcakes have been reported to increase production of phytoplankton and zooplankton which constitute the most suitable food items for carp hatchlings. But this involves problems in the procurement, transport and handling of large quantities of manures. Inorganic fertilizers, in smaller quantities, further help in enhancing the growth of phytoplankton and subsequent development of zooplankton. Therefore, a combination of organic manures and inorganic fertilizers is preferred in proper nursery management as it saves bulk handling.

Manuring is the process of application of inorganic or organic materials providing nutrients to the water as well as to the soil. Manuring boosts the plant and animal life in the nursery pond as it ultimately supply food for the fish seed.

Types of Manures and Fertilizers Used in Nursery Ponds

Organic manures

Green manures

Animal dung

Groundnut oilcake/other edible oilcakes.

Inorganic fertilizers/chemicals

Lime stone

Phosphatic fertilizers (P_2O_5)

Potassium fertilizers (K_2O)

Nitrogenous fertilizers

Cobalt chloride

Recommended Doses of Manures and Fertilizers

Organic Manures

Green manure: Grass or mangrove leaves: These are applied @ 1775 kg/ha.

Animal dung:

Cow dung: It is applied @ 10,000 kg/ha, two-three days before the spawn is stocked and @ 5000 kg/ha, 7 days after spawn has been stocked.

Poultry droppings: These are applied @ 1,000 kg/ha, 10 days before spawn is stocked and 500 kg/ha, 7 days after the spawn has been stocked.

Pig dung or house dung: It is applied @ 800 kg/ha, 10 days before spawn is stocked and 400 kg/ha, 7 days after spawn has been stocked.

Groundnut oil cake: It is applied @ 1000 kg/ha.

Inorganic Fertilizers/Chemicals

Lime stone: Lime stone is a fertilizer material in which calcium as one of the essential plant nutrients. It is applied @ 250 kg/ha.

Phosphatic fertilizers (P_2O_5): The important phosphatic fertilizers which are commonly used in the nursery pond are single super phosphate and diammonium phosphate. The effect of phsophatic fertilizer application lasts for 2 to 3 years. These are applied @ 50kg/ha.

Potassium fertilizers (K_2O): Potash is also one of the essential nutrients. It can be made available through Muriate of potash and potassium sulphate. These are generally applied along with other fertilizer materials @ 30 kg/ha.

Nitrogenous fertilizers: Nitrogenous fertilizers often have a stimulating effect on the growth of phytoplanktons which are one of the most important raw materials for fish production in ponds. The main source of nitrogen is urea or ammonia sulphate. It is applied @ 80 kg/ha.

Cobalt chloride: It is applied @ 2kg/ha. Popular combination of manures and fertilizers used in nursery pond management: The combination is: (i) Single manuring: Raw cow dung 5,000 kg + groundnut oilcake 250 kg + superphospahte 25 kg + nitrogenous fertilizer 25 kg/ha.

Phased Manuring

One day proceeding spawn stocking, superphosphatic lime 50 kg + cow dung or cattle dung 700 kg + oilcake 700 kg/ha.

On the 2nd day of stocking: 350 kg oilcake + 87.5kg cattle dung/ha.

On the 3rd day of stocking: 175 kg oilcake + 43.75 kg cattle dung/ha.

On the 4th to 9th day of stocking: Cattle dung 87.5 kg + oilcake 100 kg + superphospahte 22 kg/ha.

Oil Emulsion in the Nursery Pond

Due to manuring of ponds large number of aquatic insects develop in it. The main insects are swimmers (Notonecticdae), water boatman (Family: Corixidae), beetles, larvae and bugs. Some of the insects also fly from pond to pond. As these insects prey upon the spawn, the survival of the spawn is reduced; it is, therefore, essential that the pond is cleared of the insect populations immediately before stocking of spawn and sometimes also about a week after stocking. The aquatic insects come to the water surface very often to breathe oxygen from the air through their tracheal tubes. The best method to kill these insects is by spreading an oil film on the surface of the water. Due to this operation, the tracheal tubes of the insects get chocked with the oil film and they are unable to breathe in oxygen from the atmosphere, and hence they die.

Methods of Application

Oil emulsion is generally applied by the following two ways:

Mustard or coconut oil and cheap washing soap in the ratio of 56: 18 kg/ha.

Light speed diesel oil 50 l/ha mixed with 1/3rd to 1/4th of its weight of cheap soap.

Oil or light speed diesel oil is put into a bucket, and soap and water are added to it. Then it is thoroughly stirred till the entire solution becomes light brown in colour. It is then sprinkled on the water surface of the pond. After one hour, the insects die. The dead insects should be sieved out from the pond with a fine meshed muslin/nylon cloth of the size and depth of the pond so that all the dead insects are removed in one or two operations.

Table 9.1: Manuring of Nursery Ponds

Lime Stone (Kg)	P_2O_2 (Kg)	Urea (Kg)	$Ca(NO_3)_2$ (Kg)	K_2O (Kg)	Cow Dung (Kg)	Oil Cake (Kg)
250	25	—	—	20	—	—
—	20	—	20	—	5,000	250
—	20	—	20	—	5,000	—
350	—	—	15	—	—	—
400	—	—	—	15	—	—
250	—	—	15	10	—	—
250	—	—	15	10	—	—
300	—	15	—	10		

Precautions

Remove all the unwanted odours from the manures.

Use the first dose of manure/fertilizer or mixture, 10-15 days before the spawn is stocked.

Avoid over fertilization.

Avoid fouling of water after fertilization.

Check the growth of unwanted insects which is harmful to the young spawn.

Apply oil emulsion 12-24, hours before stocking of the pond in calm and still weather.

Procedure

Measure the total area of water.

Procure the previous report on pond water analysis and the quantum of animal and plant life. In case it is not available, take the sample (soil, water and plankton) for analysis.

Mix the required quantity of manures or fertilizers.

Make a paste of the fertilizer mixture by adding the required quantity of water.

Remove unwanted gases and odours from the fertilizer mixture by aeration.

Broadcast the slurry/paste homogeneously.

Apply oil emulsion 12-24 hours before stocking spawn in the pond.

Apply the remaining quantity of manures or fertilizer mixture in phases, as recommended.

Take water samples periodically and analyse them.

Observations and Calculations

The students should observe and analyse the following:

Plankton growth from day to day.

Physico-chemical conditions of the pond.

Oxygen level.

Alkalinity.

Primary production of the ponds in terms of carbon.

Periodic soil analysis.

Growth and percentage of survival of spawn.

Colour of the pond.

Combine manures and fertilizers for better utilization and optimum growth.

Work-out the phased manuring.

Chapter 10

Various Types of Fish Ponds and the Implements Used in Fish Farming

Fish culture in tanks, though on a limited scale, has been practiced in India since as early as 350 B.C. The industry apparently became more popular in later period. Tanks were regularly stocked with fish, and their fishing rights were leased out periodically. Fish culture is very popular in the states of Bihar, Bengal and Orissa. In other states too, there has been a marked awakening and interest in inland fish culture. At present, about 7, 00,000 acres of inland waters are under fish culture in our country. This area; is very small as compared to the actually cultivable inland waters. In Chennai (Tamil Nadu) alone, there are over 8, 50,000 acres of inland waters, including estuaries and backwaters. There are over 40,000 tanks of widely varying areas. At present, in Haryana State, about 3,000 acres of inlands water area are under fish culture practices. In India, as a whole, irrigation canals exceed 1, 00,000 km in length and wells are about 13.5 million in number. Ponds and tanks are more numerous in Bengal, Bihar and Orissa than in Madras or in any other States.

Cultivable waters are of two types, perennial or seasonal, and fishes can be cultivated in ponds, tanks, natural lakes, artificial reservoirs, backwaters, irrigation canals, wells, paddy fields, jheels, bheels, barrow-pits, quarry pools, swamps, and even short-season accumulation of water in shallow areas.

Layout Plan for a Typical Fish Farm

Dependable sources of water supply to ponds and soil, which will retain water, are of fundamental importance in the selection of a site for pond construction. Low lying, water logged or marshy areas are much suited for their excavation into fish ponds, but the soil, if peaty, would reduce the productivity of waters. A clay loamy soil is considered as the best from the point of view of water retention and fertility. When it is intended to rear carps in new ponds, the importance for the construction of Nursery, Rearing and Stocking fish ponds should be remembered. A small fish farm of 10 acres in total should have the area divided as follows for the construction of different types of fish ponds.

Table 10.1

Type of Ponds	*Area (Acre)*	*No of Ponds*	*Size (ft.)*	*Level of Water (ft.)*
Nursery	0.5	20	40 × 20 × 5	3.5–4.5
Rearing	2.0	8	80 × 40 × 6	4.5–5.5
Stocking	7.5	4	Any length × 100 × 8	6.0–7.5

The nursery ponds should preferably be small and shallow, with 3 to 4 feet depth of water. Rearing ponds should be slightly larger, but need not be proportionately deep. Stocking ponds should be deeper and longer (six to eight feet and 250 to 300 feet) so that the fish will have a longer run for a healthy growth. The width of the stocking pond should not exceed 100 feet, so that relatively smaller sized fishing nets and a limited number of fishermen would be sufficient to net the fish. Large ponds are not only difficult for fishing but also difficult to manage, particularly in the control of the overgrowing weeds.

Shallow waters are more productive than deeper waters. If there is a regular and well assured supply of water in the nursery ponds, it may be kept very shallow (2 to 2.5 feet), as in a rice field. The bund should, however, be properly tufted to prevent erosion. Fully protected inlets and outlets should be provided in fish ponds, wherever necessary, to regulate the inflow and outflow of water.

The side and bottom of the ponds should, gently be sloped towards the outlet of the pond, so as to facilitate the complete draining of water, whenever desired. If a deeper pit or depression is provided near the outlet in the pond, all the fish in pond would automatically be collected there, when the pond is completely drained.

Various Types of Fish Equipments and Fishery Methods

Primitive man was entirely dependent on the wild life existing around him to fulfill his daily needs. He has been using various methods to catch fish since the pre-historic times, and the fishing gear has undergone evaluation in different parts of the world giving rise to various methods of the present day. The method in use at a particular place depends on the type of fish caught, nature of the river and available capital. The amount of the catch by a unit depends upon its efficiency and the productivity of the fishing grounds. Most of the fishing gears have to break off operations after a certain period of activity, for rest and repair. The entire period between the launching of a fishing gear and launching it again after a gap is called a fishing cycle. The number of fishing cycles depend upon the daily pattern of occurrence of fishes, type of fishing methods used, geographical conditions of fishing grounds and the fisherman. It is difficult to enumerate all the fishing methods in use in various parts of India and there is no uniformily in the names given to various fishing gears used in commercial fisheries. Some of the common methods are described below:

Fishing Gears

Any device used for catching fish on commercial basis is called a fishing gear. There may be grouped into 4 types: nets, hooks and lines, trap and spears and harpoons on the basis of their form and mode of action. About 96 per cent of the world fishing is done with the nets.

Fishing Without Gear

This is the simplest method for catching the fish, which are caught by hand in shallow water in some hilly parts or near dams, water flows out through narrow channels, the fishes that come out in shallow channels are picked up by hand. Mollusca, pearl-oysters and murrels are also caught by his method.

Wounding Gear

In this method the fishes are caught by weapons like spear, harpoon, lance and lifts. This type of weapons are used for small area. A spear consists of a split bamboo shaft 1.50 to 2 meter in length, fitted with a conical iron point. It is thrown at a large fish by a man standing in the boat. It is also used to spear the fish through holes in the ice, or while the fish rest in streams. In this method large size carps and catfish are caught.

Stupefying Method

In this method poisons are added in the pond water and the fish is paralyzed. But this method is not beneficial. When we use poisons, is all the fishes in the water are destroyed. Certain chemicals affecting the central nervous system are also used. These are obtained from the plant *Devis eliptica, Derris uligmora* and are useful fish poisons.

Line Fishing

This method has been in use for a long time. The principal of line fishing is to offer real or artificial bait to entire the fish, which is then unable to release the bait and is lifted from the water. Previously, a thorn was used as a hook, but now metallic hooks of various shape and sizes are used. The simplest form of this gear is the hand line which consists of one mole hook attached at the end of a cotton line holding hand. Some people use a long bamboo stick at the end of which a thread and hook is tied.

Fishing by Baited Springs

A strong but flexible piece of bamboo pointed at both the ends is used as a spring. Its two ends are bent till they nearly meet and are carefully adjusted within the body of a grass hopper, cockroach or a small frog serving as a bait. This is then suspended in water by means of a string to which a float is also attached. When the fish seizes the bait, two ends of the spring separate in the month. It is used for catching murrels.

Fish Screening

Bamboo sticks are woven to form a screen of about 10 meter length and 1-1.5 meter width. The screen is set to surround a shallow tidal area at the time of high floods. When the flood water recedes the fish are left behind in the muddy pool and are collected. The screen is also set to mark off shallow areas of rivers. The enclosed space is then divided into smaller enclosures by putting up earthen bunds. The water is then drained out so as to expose the fishes. This method is employed for catching carps, catfishes, murrels etc.

Nets

A net is strong webbing in which the twins are loosely interlaced leaving regular meshes in between. The twins may be formed of a natural product or a synthetic material.

Accessories for Nets

Certain accessories are required for using the fishing net. These include ropes and cables, stakes, floats, snibers, anchors, thimbles, swives, shackles and snoods.

1. *Ropes and cables*: There are used for spreading, hanging, laying, assembling and hauling of net. They may be formed of cotton, flax, hemp, nylon etc.
2. *Stakes*: There are wooden or iron parts sharpened at one end for driving into the ground. They are used to mark spot or boundary.
3. *Floats*: These are used to keep a pond of the net floating. They may be wooden, cork or plastic blocks or empty water proof covers.
4. *Sinbers*: There are used to keep a part of the net down in water. They may be cast iron load or simply sloves.
5. *Anchors*: There are used to hold a large net in place. They are cast iron bars with a single or double fluke at one end.
6. *Thimbles*: There are galvanized steel sheet used to protect the ends of cables.
7. *Swivels*: These are steel devices used for allowing movement of ropes or cables.
8. *Snackles*: There are coupling devices used in assembling.
9. *Snoods*: These are additional twins used to the pull on the twins of the hanging nets.

Types of Nets

The principal types of nets used for fishing in the world are listed below:

1. *Active netting*: Seines, Trawls, Drag nets
2. *Passive netting*: Gillnets, trap nets
3. *Miscellaneous*: Dip nets, casting nets, hand nets

Seines, gill nets and trap nets are used for about 56 per cent, 20 per cent and 20 per cent of wild fishing, respectively.

Active Netting

The net has (a) float linear hard rope bearing a number of floats and (b) a load line or fort rope carrying a number of sinbers.

Seine

Seine is a very large net used for fishing in lakes, river and sea. It may be 1000 m long, 12 m wide and about 5500 kg in weight. It is very strong to withstand (*i*) the pull of hauling (*ii*) the resistance of water to its movement (*iii*) the weight of captured fish. Its mesh size should be such that does not 'kill' the fish.

Structure

A common beach seine consists of two wings joined to a cam bed bag all formed of net-webbing. The wings may be equal or unequal in size, but usually taper towards the ends. The ends are joined to the wraps directly or through brail. The float line and load line bear floats and sinbers respectively and have spreaders extending between them at regular intervals. The mesh size of the wings is larger than that of the bag.

Trawls

Trawls are large nets used for fishing in the mild water and near the bottom all over the world. It is a conical bag of net webbing with a wide mouth and a tapering blind code end. Mouth bears a float line along the upper margin and a foot rope along the lower border. The flat line serves to keep the mouth open. The foot rope is fitted with wooden, metallic or plastic bobbins to prevent wear and tear of the net during dragging. Beyond the mouth the various lines extend as bare ends referred to as legs of lines. Between the legs of lines and the warps are special fittings which include other board, sweep lines and spreaders. The warps of a trawl are 200m long. The trawl is dragged by one vessel or by two vessels.

Drag Nets

The drag net, locally called pattijal is a rectangular piece of net webbing. Its upper margin carries a float line with floats and the lower margin bears a load line with weights. This arrangement helps to keep the net in a vertical position in water. The net is dragged along the bottom of the pond by a team of men. The latter wrap through the pond, shouting and splashing water, and driving the fish ahead. The fish are driven to one side of the pond where they are trapped in the net.

Miscellaneous

Dip Nets

A dip net consists of net webbing mounted on a v-shaped rigid frame. Dip nets are of several shapes and are lowered into the water in the hope that fish would swim over them and lifted out of water. They are triangular, rectangular in shape and are fitted round a flame. Some bait is often put on the net or suspended over it to attract fishes. The number net is operated by hand, while the longer ones are dipped and lifted from the water by means of a long pole which is operated like a lever. The digital simple hand dip nets have gradually been transformed into permanent stationary lift nets operated from bridges made of bamboos.

Cast Net

It is a circular net having the shape of a large umbrella. A strong card is attached to the apex of the umbrella and a number of load or iron weights are fixed all along the margin. The fisherman throws the net fully spread over the water, keeping the long rope in his left hand. This has to be done very skillfully so that the net falls on the surface of the water fully expanded. The net sinks to the bottom and the circumference clocks due to the weights attached to it. All kinds of small sized fish are entangled in the net, which is then pulled out by means of the cord. The cast net is extensively used in ponds and rivers, and all along the sea coast, but can not be used in places full of weeds or with rocky bottom.

Hand Net

The hand net is a net cloth bag with 4 to 6 mm wide meshes. It measures 70 to 75 cm in length and 50 to 55 cm in width. Its open end is stitched around an oval frame of bamboo, cane or metal. The frame is fitted with a handle on one side to operate the net. The hand net is used to transfer the breeders from stocking ponds to breeding ponds and hapas and back. It is also used for weighing the fish and for holding it for pituitary extract injection.

Mesh Size

Mesh sizes recommended for handling the fish of various sizes and age groups are given in Table 10.2.

Table 10.2

Sl.No.	*Fish Size*	*Mesh Size*
1.	Fry (4–5 cm)	4–6 mm
2.	Fingerlings (10–15 cm)	8–10 mm
3.	Fingerlings (25–30 cm)	2.5–5 cm
4.	Adults (40–50 cm)	5 cm

Chapter 11

Various Aspects of Management Practices Used in Fish Farming

(A) Liming and pH Adjustment in Nursery Ponds

What are the important liming materials used in nursery pond management?

Generally calcium oxide (CaO) called slaked lime or quick lime and calcium carbonate ($CaCO_3$) called chalk or lime stone are used as the liming material.

Why liming is done in nursery pond?

It is done to supply calcium which is an important nutrient. It not only acts as a buffer but also keeps the pH value on the alkaline side. By its application, acidic agents are neutralized. Moreover, it also acts as a disinfectant and kills harmful bacteria, and keeps the water clean. It can be used for the control of parasitic fish diseases also.

It also releases certain nutrients like potash and phosphate from the bottom sludge into the water which go into the production cycle.

The dosage of lime is calculated, based on the prevailing pH value of the pond.

pH	*Pond Condition*	*Dosage of Lime (kg/ha)*
4–4.5	Highly acidic	1000
4.5–5.5	Medium acidic	700
5.5–6.5	Slight acidic	500
6.5–7.5	Near neutral	200

Pond water having a pH value 7.0 is neutral. When it is less than 7.0, it is acidic and when it is more than 7.0, it is alkaline. Acidic waters are unproductive. Hence, the pH of the pond water should be maintained on the alkaline side. If the pH value of the pond water is around 7.5 to 8, the pond is considered as highly productive. But if the pond water is near neutral, further liming is done as it increases the productivity of the pond.

Precautions

Apply the liming material homogeneously.

Use liming material in powder form.

See that liming is done before fertilization.

Use hand gloves while liming.

Procedure

Find out the pH value of the pond with the help of litmus paper or pH strips.

Calculate the quantity of lime required.

Put on the rubber gloves.

Weigh the liming material and convert it into powder.

Take the powdered lime in a bucket and spread it evenly in the pond.

Wash your hands thoroughly.

Find out the pH value after liming.

(B) Manuring and Fertilization of Nursery Ponds

Aims and objectives

To understand the importance of manuring of nursery ponds.

To distinguish between different manures and fertilizers.

Relevant Information

Manuring is an important step in the management of nursery ponds for better growth in the rearing of carp spawn. Organic manures like cow dung and various oilcakes have been reported to increase production of phytoplankton and zooplankton which constitute the most suitable food items for carp hatchlings. But this involves problems in the procurement, transport and handling of large quantities of manures. Inorganic fertilizers, in smaller quantities, further help in enhancing the growth of phytoplanktons and subsequent development of zooplanktons. Therefore, a combination of organic manures and inorganic fertilizers is preferred in proper nursery management as it saves bulk handling.

Manuring is the process of application of inorganic or organic materials providing nutrients to the water as well as to the soil. Manuring boosts the plant and animal life in the nursery pond as it ultimately supply food for the fish seed.

Types of Manures and Fertilizers Used in Nursery Ponds

Organic manures

Green manures

  Animal dung

  Groundnut oilcake/other edible oilcakes.

Inorganic fertilizers/ chemicals

Lime stone

Phosphatic fertilizers (P_2O_5)

Potassium fertilizers (K_2O)

Nitrogenous fertilizers

Cobalt chloride

Recommended Doses of Manures and Fertilizers

Organic Manures

Green Manure

Grass or mangrove leaves: These are applied @ 1775 kg/ha.

Animal Dung

Cow dung: It is applied @ 10,000 kg/ha, two -three days before the spawn is stocked and @ 5000 kg/ha, 7 days after spawn has been stocked.

Poultry droppings: These are applied @ 1,000 kg/ha, 10 days before spawn is stocked and 500 kg/ha, 7 days after the spawn has been stocked.

Pig dung or house dung: It is applied @ 800 kg/ha, 10 days before spawn is stocked and 400 kg/ ha, 7 days after spawn has been stocked.

Groundnut Oil Cake

It is applied @ 1000kg/ha.

Inorganic Fertilizers/Chemicals

Lime Stone

Lime stone is a fertilizer material in which calcium as one of the essential plant nutrients. It is applied @ 250 kg/ha.

Phosphatic Fertilizers (P_2O_5)

The important phosphatic fertilizers which are commonly used in the nursery pond are single super phosphate and diammonium phosphate. The effect of phsophatic fertilizer application lasts for 2 to 3 years. These are applied @ 50kg/ha.

Potassium Fertilizers (K_2O)

Potash is also one of the essential nutrients. It can be made available through Muriate of potash and potassium sulphate. These are generally applied along with other fertilizer materials @ 30 kg / ha.

Nitrogenous Fertilizers

Nitrogenous fertilizers often have a stimulating effect on the growth of phytoplankton which is one of the most important raw materials for fish production in ponds. The main source of nitrogen is urea or ammonia sulphate. It is applied @ 80 kg/ha.

Cobalt Chloride

It is applied @ 2kg/ha. Popular combination of manures and fertilizers used in nursery pond management: The combination are: (i) Single manuring: Raw cow dung 5,000 kg + groundnut oilcake 250 kg + superphosphate 25 kg + nitrogenous fertilizer 25 kg/ha.

Phased Manuring

One day proceeding spawn stocking, superphosphatic lime 50 kg + cow dung or cattle dung 700 kg + oilcake 700 kg/ha.

On the 2nd day of stocking: 350 kg oilcake + 87.5kg cattle dung/ha.

On the 3rd day of stocking: 175 kg oilcake + 43.75 kg cattle dung / ha.

On the 4th to 9th day of stocking: Cattle dung 87.5 kg + oilcake 100 kg + superphospahte 22 kg /ha.

Oil Emulsion in the Nursery Pond

Due to manuring of ponds large number of aquatic insects develop in it. The main insects are swimmers (Notonecticdae), water boatman (Family: Corixidae), beetles, larvae and bugs. Some of the insects also fly from pond to pond. As these insects prey upon the spawn, the survival of the spawn is reduced; it is, therefore, essential that the pond is cleared of the insect populations immediately before stocking of spawn and sometimes also about a week after stocking. The aquatic insects come to the water surface very often to breathe oxygen from the air through their tracheal tubes. The best method to kill these insects is by spreading an oil film on the surface of the water. Due to this operation, the tracheal tubes of the insects get chocked with the oil film and they are unable to breathe in oxygen from the atmosphere, and hence they die.

Methods of Application

Oil emulsion is generally applied by the following two ways:

Mustard or coconut oil and cheap washing soap in the ratio of 56: 18 kg/ha.

Light speed diesel oil 50 l/ha mixed with 1/3rd to 1/4th of its weight of cheap soap.

Oil or light speed diesel oil is put into a bucket, and a soap and water is added to it. Then it is thoroughly stirred till the entire solution becomes light brown in colour. It is then sprinkled on the water surface of the pond. After one hour, the insects die. The dead insects should be sieved out from the pond with a fine meshed muslin/nylon cloth of the size and depth of the pond so that all the dead insects are removed in one or two operations.

Table 11.1: Manuring of Nursery Ponds

Lime Stone (Kg)	P_2O_2 (Kg)	Urea (Kg)	$Ca(No_3)_2$ (Kg)	K_2O (Kg)	Cow Dung (Kg)	Oil Cake (Kg)
250	25	–	–	20	–	–
–	20	–	20	–	5,000	250
–	20	–	20	–	5,000	–
350	–	–	15	–	–	–
400	–	–	–	15	–	–
250	–	–	15	10	–	–
250	–	–	15	10	–	–
300		15	–	10		

Precautions

Remove all the unwanted odours from the manures.

Use the first dose of manure/fertilizer or mixture, 10-15 days before the spawn is stocked.

Avoid over fertilization.

Avoid fouling of water after fertilization.

Check the growth of unwanted insect life which is harmful to the young spawn.

Apply oil emulsion 12-24, hours before stocking of the pond in calm and still weather.

Procedure

Measure the total area of water.

Procure the previously done report on water analysis and the quantum of animal and plant life. In case it is not available, take the sample (soil, water and plankton) for analysis.

Mix the required quantity of manures or fertilizers.

Make a paste of the fertilizer mixture by adding the required quantity of water.

Remove unwanted gases and odours from the fertilizer mixture by aeration.

Broadcast the slurry/paste homogeneously.

Apply oil emulsion 12-24 hours before stocking spawn in the pond.

Apply the remaining quantity of manures of fertilizer mixture in phases, as recommended.

Take water samples periodically and analyse them.

Observation and Calculations

The students should observe and analyse the following:

Plankton growth from day to day.

Physico-chemical conditions of the pond.

Oxygen level.

Alkalinity.

Primary production of the ponds in terms of carbon.

Periodic soil analysis.

Growth and percentage of survival of spawn.

Colour of the pond.

Combine manures and fertilizers for better utilization and optimum growth.

Work-out the phased manuring.

Chapter 12

Common Aquatic Weeds of Freshwater Fish Ponds

Aquatic weeds are defined as those unwanted and undesirable plants which reproduce and grow in water and if left unchecked, choke the water body thus posing a serious threat to pisciculture, as they consume large quantities of soluble nutrients and drastically curtail the fish production. Weed choking of a body of water severely restricts plankton production; limits the living space of fish; upsets the equilibrium of physico-chemical qualities of water, chokes O_2 circulation, promotes accumulation of deposition of siltation; provides shelter to predatory and weed fishes, molluscs and aquatic insects and obstructs the netting operations.

Thus, weeds on the basis of their habits and habitats, having been variously classified. Lawrence (1955) grouped them into following types:

Floating Weeds (Figure 7)

Which have their free floating leaves above the surface of water with roots hanging free underneath, *e.g. Eichornia, Pistia, Spirodela, Lemna, Azolla* and *Wolffia.*

Emergent Weeds (Figure 8)

They are rooted in the bottom but having their floating leaves flush with the water surface and the stems are always hard, remain inside water body, *e.g. Nymphae (lotus), Nelumbo* and *Nymphoides.*

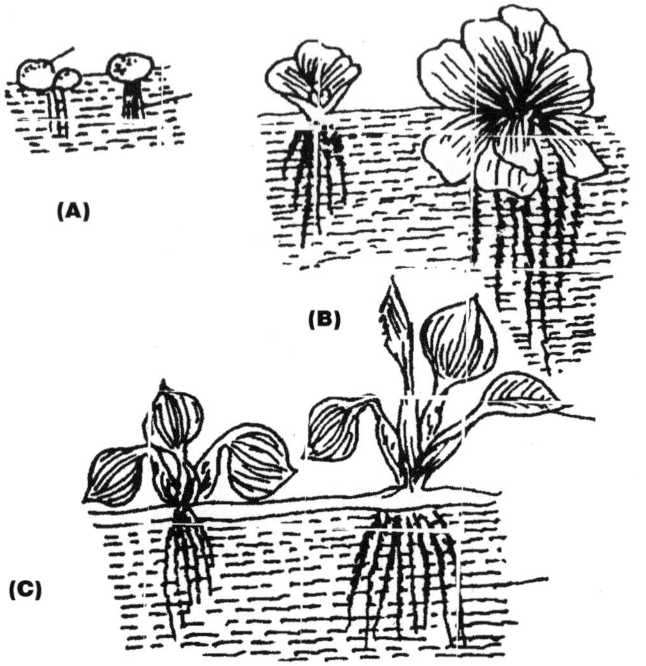

Figure 7: Floating Weeds
(A) ***Lemna*****; (B)** ***Pistia*****; (C) Waterhyacinth**

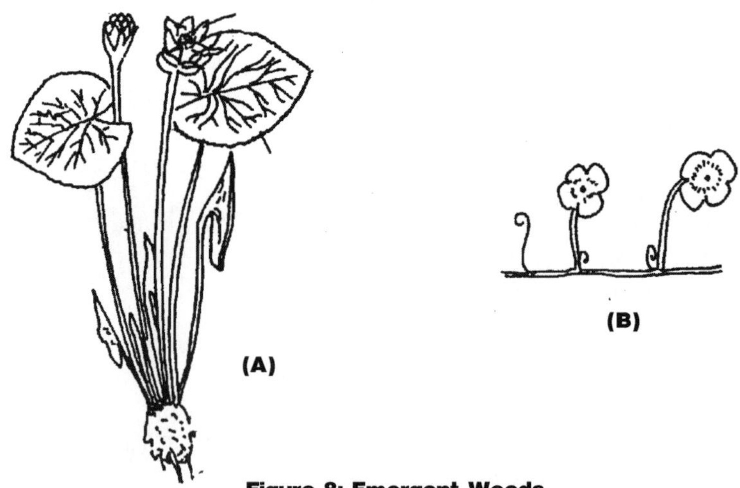

Figure 8: Emergent Weeds
(A) ***Nymphae*****; (B)** ***Nelumbo***

Submerged Weeds (Figure 9)

Which may or may not be rooted but whole of the body of the plant remains inside the water except the flowering part in some case, *e.g.* rooted *Hydrilla Najas, Vallisneria, Potamogeton, Chara* devoid of roots: *Ceretophylla, Utricularia, Najas,* etc.

Marginal Weeds

These are mostly rooted and choke the shore line of the water body, they are very hard and can susain without water for two-three months, *e.g. Typha, Pharagmities, Cyperus, Panicum, Colocasia, Ipomoea.*

Mats or Scum

They are formed by filamentous algae and some *Chlamydomonas* and Euglininae family in the marginal area or in the main body of the water, thus form a thin film over the surface of water, *e.g. Spirogyra, Pithophora,* etc.

Algae Scattered through the Water Body

In this case these micro aquatic plants give greenish tinge to water body due to their presence when in abundance. They are scattered on the surface, column and bottom of the pond water, *e.g. Mycrocystis, Oscillatoria, Anabanena,* etc.

Phytoplanktons

These are minute microscopic plants passively floating in water, often multiplying very rapidly and occurring in millions imparting a turbid green, brown or yellowish brown colour to water, such a condition is known as "*water bloom* and the plants are generally allied algae. The majority of planktonic algae in the Indian fresh-waters belong to the groups green algae (Chlorophyceae) and blue green algae (Cyanophyeae) though a number of diatoms (Bacillariophyceae) and flagellates (Euglenineae) are also commonly found.

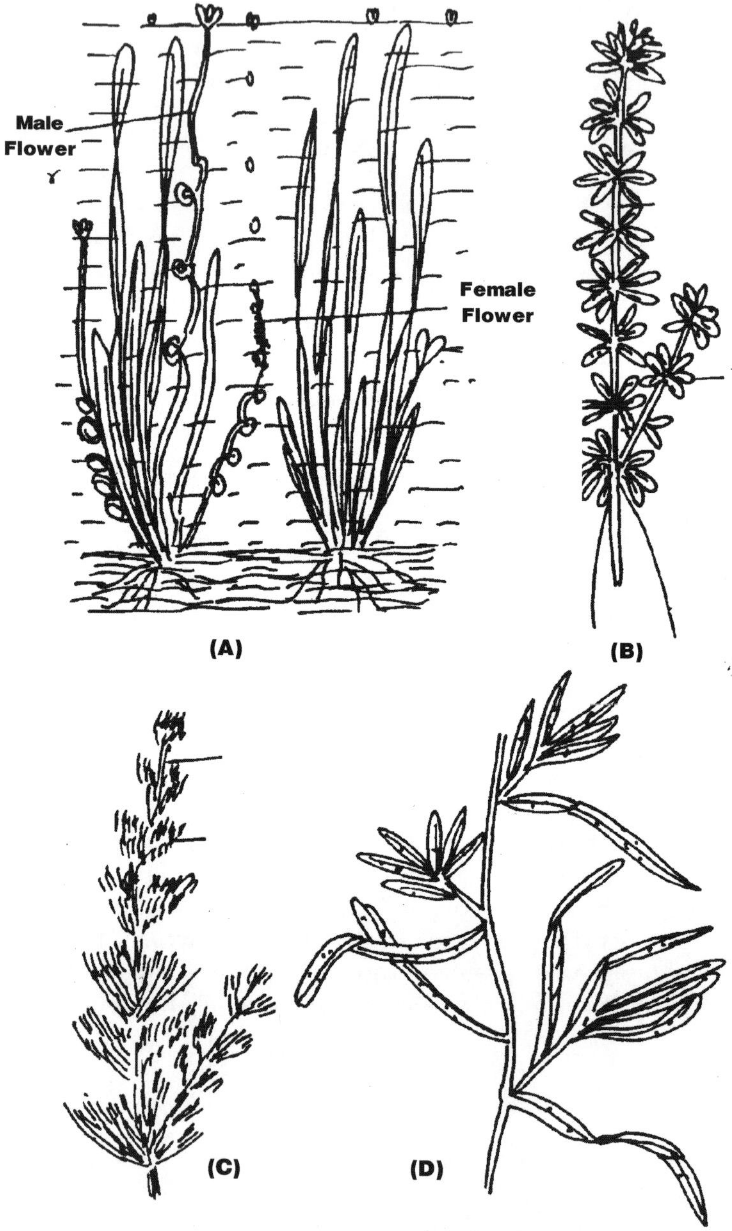

Figure 9: Submerged Weeds
(A) *Vallisneria*; (B) *Hydrilla*; (C) *Ceretophylla*; (D) *Potamogeton*

Chapter 13

Control of the Common Aquatic Weeds of Freshwater

After the pond is properly constructed, filled with water and fertilized, it supports a balanced aquatic vegetation requisite for pond ecology. When a new pond is constructed, the following precautions should be taken to check weeds from establishing themselves:

Keeping minimum depth of water at not less than 0.5 m

Keeping of the pond edge intact to reduce the hazard of marginal and shallow water weed growth.

Shaping and sodding of the pond edge above water level to reduce the area where marginal weeds could appear.

Providing a diversion or ditch, if necessary, to carry excessive dirt and muddy water around the pond. The common methods adapted for control of aquatic weeds are: manual, mechanical, chemical and biological.

Manual and Mechanical Control

Manual methods are often employed to remove weeds. In small water bodies, the traditional methods of hand picking, uprooting emergent and marginal weeds and cutting them with scythes are considered suitable. The mechanical methods are relatively slow, rather expensive. In Bengal, a weed cutting winche having its bow fitted with long V shaped sickles and having reciprocal action, have been used in cutting weeds like *Colocasia, Cyperus* and other tall weeds in shallow water, power winches or mechanical lifts have been employed in Assam to clear shallow water thick infested with water hyacinth. Recently, diesel operated winches have been used in Orissa to eradicate dense rooted submerged vegetation.

Chemical Control

A large variety of chemical weedicides are now used for the control of aquatic weeds. Their actions, either by direct contact or by translocation of the chemical from the treated part of the plant result in death of the plant. Following points should be remembered before using herbicides.

Effective in killing weeds at reasonably low rate of application, not exceeding 10 ppm, for those applied in water and about 10 kg/ha for those spread on a surface area basis.

Cheap and easily available.

Nontoxic to fish, human beings and stock animals.

Should not pollute the treated water.

Should not have an corroding action on metal container and sprayers.

Control of Floating Weeds

Among the floating weeds the water hyacinth (*Eichornia grassipers)* is most noxious and for its successful control 2, 4 diclorophenoxy acetic acid (2, 4-D) is used. It has direct effect upon the leaves and roots, if applied at the rate of 4.5–6.7 kg/ha. Other chemical is Taficide-80 (2, 4-D sodium salt 80 per cent) rate 4-6kg/ha. It shows its effect within one week without an adverse effect on fingerlings and adults of major carps.

Other floating weeds such as *Pistia stratiotes* and *Salvinia* have been controlled by H.S.D and Powerine @ 775-1100 liter/ha, MCPA (2, 4-Dester) at the rate of 8.96-13.44kg/ha at 0.2 per cent concentration which are best for killing *Pistia*.

Control of Marginal Weeds

Young *Cyperus* are totally killed with 1 per cent aqueous solution of both Taficide-80 and Tafapon (2.2-dichloropropionic acid sodium salt) at the rate of 28 kg/ha. *Elocharis* is controlled with 2.5 per cent aqueous solution of 2,4-D with detergent at 0.25 per cent concentration and *Colocasia* with Taficide-80 at a concentration of 1.5 per cent with detergent 'surf'.

Control of Emergent Weeds

In India, encouraging results have been achieved in the control of *Nymphae* with Dicotox and *Nelumbo* with Taficide-80 at a concentration of 1.5 per cent application of 2,4-D (sodium salt 80 per cent) at 1.5 per cent concentration with detergent surf. Taicide-10 at concentration 2.5 per cent with detergent showed good result on *nymphoides*.

Control of Submerged Weeds

Copper sulphate ($CuSO_4$) with ammonium sulphate (NH_2SO_4) is applied in controlling the Chara. Complete killing of *Hydrilla* and *Najas* are eliminated by using anhydrous ammonia. The gas injected directly in the top layer of the weeds chocked water by means of suitable application at the rate 1.2 ppm. Other weedicides are Simazine at about 3-5 ppm active ingredient, also sulphur dioxide at about 40 ppm was found effective in killing *Hydrilla verticillate*.

Control of Algal Blooms and Filamentous Algae

The blue algae particularly *Microcystis* is most noxious algae. For this, Simazine was experimented at the rate of 0.5-1.0 ppm active ingredient, which could effectively control *Microcstis*. This weedicide is not toxic to fish or zooplankton.

Biological Control

Certain types of weeds in a body of water can be controlled by means of selected varieties of herbivorous fishes. The more important herbivorous fishes are grasscarp *(Ctenopharygodon idella)*, Common carp, Tilapia. Gourami, an ideal herbivorous fish is one which consumes variety of weeds. Tilapia is best for *Chara* and *Najas*. Grass carp was found consuming wide variety of the aquatic vegetation.

Other animals such as duck and geese can be employed for controlling aquatic vegetation. In addition, they manure the ponds with their excreta, the geese keep the grass on pond banks trimmed short.

Chapter 14

Identification and Control of Aquatic Insects of Fish Ponds

Nursery ponds are generally abundant in aquatic insects, most of which are harmful to the fish, either as predators or competitors. Hence for the maintenance of healthy nursery ponds, it is necessary to identify and eradicate these harmful insects.

Insects belong to the phylum Arthropoda which are characterized by the presence of jointed appendages (legs, antenna, antennules, etc.)

The major characteristics of class Insecta is the division of the body into head, thorax and abdomen, presence of 2 pairs of wings and 3 pairs of walking legs. Respiration in this class of animals is by simple "air tubes" called "Trachae" traversing the entire length and breadth of the body and opening out by minute openings called "spiracles". The most commonly found freshwater insects are *Nepa, Corixa, Notonecta, Belostoma, Cybister, Gerris, Ranatra* and Dragonfly nymphs.

Precautions

Dilute formalin to 4 per cent for preservation.

Do not keep hands in formalin for long. It is better to use hand gloves while examining preserved specimens.

After examination of preserved specimens, wash hands thoroughly with detergent soap.

Materials Required

Chat jal (closed meshed drag net)

Specimen bottles, watch glasses, glass slides, trays

Dissection box

Stage microscope

4 per cent formalin

Sticker labels.

Procedure

Net out nursery pond with chat jaal and collect the insect sample.

Preserve the sample in 4 per cent formalin and take it to the laboratory for detailed examination.

Transfer the sample to the tray and separate the different insects.

Transfer the different insects to different specimen bottles and preserve in 4 per cent formalin.

Undertake detailed examination under the stage microscope to identify the different insect, using the identification key given below.

Count the number of insects in each group and find out the total count.

Key to Identification of Aquatic Insects (Figure 10)

Nepa

Body blackish and dull in colour

Head down ward in praying position.

First pair of legs held in a raptorial position.

Two long respirator filaments or breathing tubes at the tips of the abdomen.

First pair of wings covering the body.

Thorax very flattened and prothorax narrow.

Nepa commonly known as the water scorpion. This insect grows up to 17 mm in length. Because of their predatory nature, these insects are harmful to the fish fry.

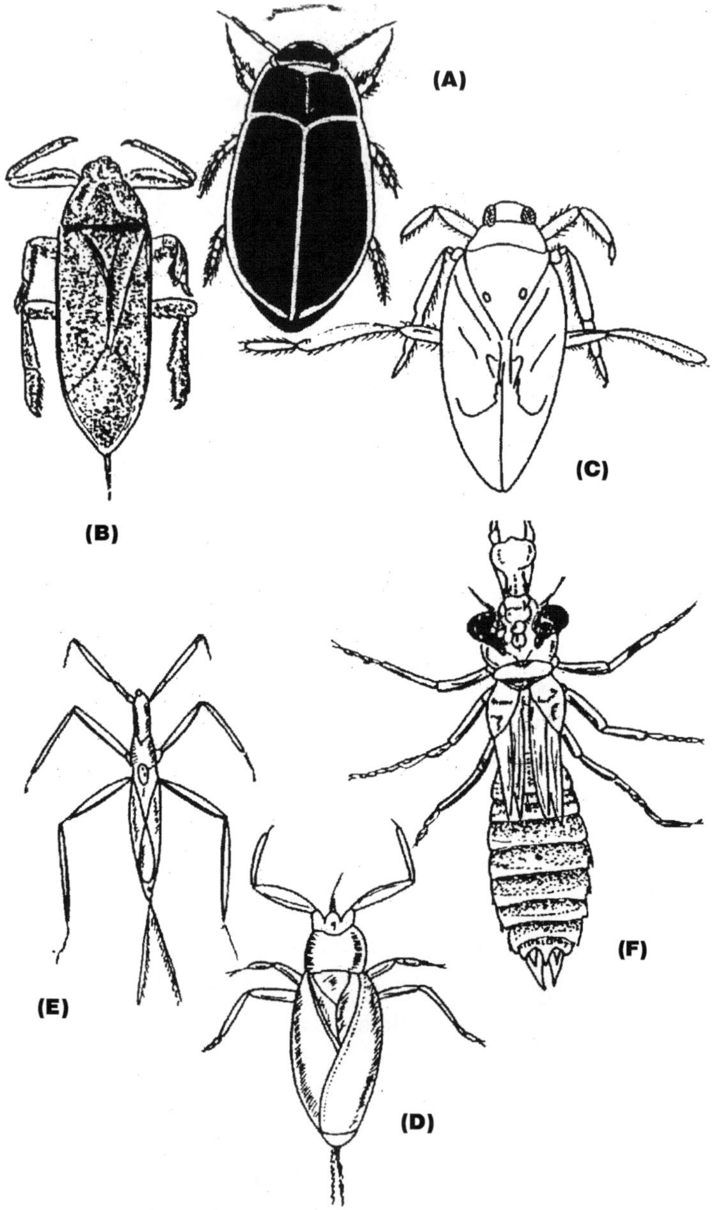

Figure 10: Predatory Aquatic Insects
(A) *Cybister* (Divine Beetle); (B) *Belostoma* (Giant Water Bug); (C) *Notonecta* (Back Swimmer); (D) *Nepa* (Water Scorpion); (E) *Ranantra* (Water Stick Insect); (F) Dragon Fly Nymph

Corixa

Body dark greyish with faint yellow cross lines.

Middle legs extraordinarily long.

Hind legs flattened and fringed.

Corixa commonly known as "water boatmen". They are very common in shallow ponds. They do not catch fry but they compete with them for food items like zooplankton and debris.

Notonecta

Body bluish black in colour.

Body deep, convex dorsally and keeled ventrally with sides sloping.

Hind legs long and fringed for swimming, swim jerkily.

Body has only 3 pairs of legs.

Notonecta is commonly known as "back swimmers". These insects grow up to 10 mm in length and are harmful to spawn and fry in the carp nursery.

Belostoma

Body brown or dull green in colour.

Body large, flat and oval.

Head bears a powerful beak.

Forelegs chelate and hind legs long and modified for swimming.

Wing functional.

Belostoma is commonly called "giant water bug". It grows to between 20-70 mm in length. These insects are predators in habit, attacking fish fry even larger than themselves.

Cybister

Shining black or brownish black body, dull hollow margins.

Head broad tightly fitting and capable of only slight movement.

Antennae small and slender

Hind legs long with hairs

Wings functional.

Cybister is commonly called "diving beetle". It grows to 35 mm and has the biting type of mouth parts.

Cybister Larva

Body muddy green in colour.

Body long slender and segmented.

Swims by up and down movements of the body. It can be seen hanging from the water surface film.

Cybister larva is commonly known as "water tigers" or "dragons of the ponds".

Gerris

Gregarious (live in group)

Head small and body slender

Forelegs small while last two pairs of legs 2-3 times longer than the body live on the water surface.

The specimen is *Gerris* or "Water spider". They grow to 2-15 mm length

Ranatra

Body dull green in colour.

Prothorax narrower than head.

Forelegs long and modified for grasping the prey.

Two long filamentous respiratory tubes at the posterior end.

Generally swims above the water surface.

The specimen in *Ranatra* or "water stick insect". It grows to 20-40 mm in length exclusive of the tracheal tube. Ranatra are very common insects in nursery ponds. The prey upon fish fry even up to 25 mm size.

Dragon Fly Nymph

Body muddy brown in colour.

Head broad, the lower side concealed by the development of the lower lip.

Antenna short.

Dragon fly nymphs are predatory in habit, consuming fish spawn and fries as well as other insects.

Chapter 15

Heavy Metal Toxicity in Fishes

Water pollution today has become world's most serious problem, as water although an essential commodity for sustaining life, is being maximally exploited by the industrialists, civic bodies vis-à-vis, the water surface transporting units. Modern civilization with its rapidly growing industrial units and ever increasing population has led to an accelerated degradation of the freshwater resources. Today about 100 million people are known to severely suffer from these polluted waters, since a variety of causative agents are known to occur for various physiological disorders and water born diseases.

Water a valuable God's gift to man, is required in healthy state for every organism. Analysis of most of the natural water bodies revealed a critical loss of biodiversity, frequent occurrence of pathogens and a variety of other noxious problems with aquatic plants and animals due to man's careless activity. In India textile and dying industries alone contribute 10 to 15 per cent of about 1000 different chemicals in nature, leading to increased BOD and COD contents, thus simultaneously decreasing DO and creating anaerobic conditions in the rivers. In Delhi about 200 million liters of untreated water is pumped into Yamuna. The major industries causing water pollution are the paint and the detergent industries, metal and surgical industries, steel and pulp industries and various other chemical industries. Fly ash is the major contributor of heavy metal toxicity in natural waters. Over 100 million tones of fly ash are produced each year and much of it is transported in the atmosphere before deposition. It is a major source of contamination of heavy metal such as cadmium in the rivers and estuarine sediments in the vicinity of such power stations. Even most of the wetlands in India including Uttar Pradesh today have become heavily polluted and unproductive.

The US Department of Health Education and Social Welfare has classified water pollutants in eight categories *viz.* (1) Sewage wastes, (2) Infectious agents, (3) Plant nutrients, (4) Particulates, (5) Radioactive substances, (6) Mineral and chemical substances, (7) Heat, (8) Organic chemical exotics. Heavy metals are one major component among various effluents which has attained alarming status as water pollutants over the last few decades due to heavy industrialization. Toxic metals are added in aquatic system from industrial processes, domestic sewage, discharge, steel dust, land runoff and fossil fuel burning. Traces of heavy metals such as Hg, Cd, Pb, As, Co, Mn, Fe and Cr have been identified as deleterious to aquatic ecosystem and human health. Excess inputs are man made is coming from industrial and mining effluents (Cd, Cu, Fe, Hg, Pb, Zn) pesticides (Cu, Hg, Zn), the burning of fuels and car exhausts (As, Ni, Pb, V, Zn). Since the rivers has become the natural dustbins for discharging all sorts of city, industrial and other wastes, these destroy natural harmony between biotic and abiotic components of the aquatic ecosystem. This envisages a great concern with the ecology of polluted waters and its perspectives with regard to its fitness for biotic community including fisheries. Later being a natural food source for about half of the population of the world needs serious implications as the polluted food could draw not only nutritional complications in terms of diseases or other physiological disorders in humans and fish production, but also produce more hazards to the living world on account of a variety of indirect interacting factors.

Mercury is an unique pollutant because of its apparent indestructibility and ubiquitousness. It attracted worldwide attention as a potential pollutant because of a tragic episode which occurred in the Japanese villages surrounding the Minamata Bay. The victims were struck with a mysterious nervous illness commonly called "Minamata disease" which was later shown to be caused by consuming fish and shell fish that has lethal accumulation of methyl mercury. It resulted in 121 cases of poisoning and 41 death. Another similar incident of mercury poisoning was reported again in Japanese village near Niigata prefecture during 1969, which left 6 people dead with 47 poisoned cases.

The chlor-alkali industry in India has grown ten fold since 1947. Large quantities of caustic soda and soda ash produced are used in Paper and Rayon industries. Majority of the chlor-alkali factories adopt old process of using mercury as cathode. This has long since been discarded in Japan and in many other countries. The country imports around 200 tons of mercury every year of which 180 tons are used by the chlor-alkali industry alone and most of it finally goes out along with waste water and sludge, and ultimately enters into the environment.

Central Inland Fisheries Research Institute and National Environment Engineering Research Institute in 1981 reported that Hooghly estuary received 28.5 ppm of chromium from tannery and 0.007 ppm to 129.7 ppm of mercury from other various industries. Kaviraj (1989) also surveyed Hooghly estuary during 1977-81 and reported that an accumulation of 381.6 ppm of zinc and 9.6 ppm of chromium in the sediments, 120.4 ppm of zinc, 127.8 ppm copper and 20 ppm chromium in fishes and 112.2 ppm zinc, 10 ppm copper and 0.3 ppm chromium in shrimps.

The realization of the deleterious conditions of the rivers in the early nineteen century in Britain on account of disappearance of salmon from the Thomei and Mersey rivers prompted passing of certain acts to curb water pollution. In 1857, a royal commission on prevention of river pollution was set up. The federal water pollution central act of 1948 passed in America ascertained water quality and a new act passed in 1965 required the states to establish water quality standards. In India, the water (prevention and control of pollution) act came into existence in 1974 only and thereafter the environmental act was passed on 23 May, 1986. As per WHO, permissible limits for toxic metals in drinking water are: cadmium, 0.01, lead 0.1 and chromium 0.05 mg per litre. As per the analysis of ground water quality as a measure of impact of Bhilai steel plant effluents, it has been reported that the post monsoon ground water samples representing entire water shed area are in general high in toxic elements.

Two kind of heavy metal occur in a animal body, one termed as essential metals and others are non-essential metals. Non-essential metals are present in intracellular compartment and appear toxic to various animal function because of their binding to metallothione (MT). Experimental studies have proved that the role of heavy metal ions in living systems follows the pattern of natural availability. There is no life, that can survive without the participation of metal ions. Some of the major ions such as Na, K, Mg, Ca are essential to sustain life. In the same way some of the heavy metals are considered essential both for plants and animal nutrition, and they serve some useful biological functions in the body. Thus, Cu, Co, Mn, Mo, Se and Zn are essential to both plants and animals. However, some metals such as Cd, Hg, Pb and metalloids like As, Sb and Se are considered to be toxic, although some essential physiological roles have been inferred for As, Cd and Pb recently (Dara, 1993). Among the metals studied Cr, Ni, Cd and Be have been found to be potentially carcinogenic (*i.e.* causing cancer). Even the heavy metal although essential, on "acute exposure" or "chronic exposure", severely affect different body organs and produces various types of diseases and deformities in the human body.

Copper necessary for the normal biological activity, excessive intake however, can be toxic and can cause haemolysis and a disease of liver called Willson's disease. Generally liver, kidney, gills and intestine are the main site for MT formation in fishes which is either stored or excreted. Heavy metal residues, nickle, chromium, cobalt in combined waste water from vegetable oil industries are generally below detectable limits (< 0.01 mg/l). These waters however, are reported to be iron rich (Av. 1.5 mg/litre), and implying very poor DO levels.

According to Haslan (1978) the effects of sewage and industrial affluent on aquatic biotic system can be described as:

1. A decrease and loss of the species, most sensitive to pollution.
2. A decrease in species diversity at the sight of affluent industries.
3. An increase in any species actually favoured by the polluted waters.

Heavy metals are important, not only as a source of phytotoxicity in plants but also these enter into food chain causing hazardous impact on animals, including the humans. Chromium accumulation in an animal body occur mainly in liver, whereas lead and cadmium accumulation do not vary significantly in different tissues. The major symptoms of heavy metal toxicity in mammals are hypertension, cirrhosis of liver, lung cancer, nausea, vomiting, salivation, diarrhoea, abdominal cramps etc.

Body size is closely related to fish growth and metabolism and also an important attribute to metal accumulation in fishes. Liver and kidney mainly act for storage and transformation of such xenobiotics (Heath, 1995). Maiti and Banerjee (2002) studied bioaccumulation of certain heavy metals in three fish species such as *Labeo rohita, Catla catla* and *Oreochromis nilotica* in wastewater fed wet lands in Calcutta. They reported considerable role of the age of fish in metal accumulation in different fish species. The bioaccumulation of metals in tissues is further species specific.

The expression 'uptake efficiency' has been used to refer to how readily a fish removes a contaminant from the water by the gills. The metal uptake is however, greatly influenced by pH and temperature. Copper uptake by the gills of feathered minnows were found to the slower at low pH, cadmium appear to be taken up in direct correlation with pH and lead exhibit exactly the opposite *i.e.* the pH increases uptake (Camphell and Stokes, 1985). In terms of biomagnifications however, different metals behave differently for instance, mercury biomagnifies easily, whereas cadmium and lead seemingly do not, nor do the body concentration of the later two metals increase with age and size of the fish, as dose mercury. The distribution of metals into body tissues and organs appear to very greatly depending upon the fish species. Lead accumulates mainly in scales, gills, bone and liver and in a estuarine fish liver and muscles accumulated almost no lead (Spry and Weiner, 1991).

Relationship between mercury content and weight of the fish and fish length has been reported by many workers. They found the mercury content to increase with increase in the body weight. Llyod (1977) measured Hg levels in the liver and muscle tissues of rainbow trout and brown trout from trout parks stream of southern Missouri and found a relationship between concentration and size. Few studies showed decrease in the mercury level with increase in the size.

Tariq *et al.* (1979) reported increased accumulation of mercury in muscle and liver with many cases of increased level in muscle than liver. Out of several species of fishes analysed by Greig *et al.* (1975) most of the species were found to contain equal amount of mercury in muscle than in liver and a few cases of increased level in liver than in muscle.

Acute levels of metals in water, however, is said to stimulate mucus secretion which serve as a way of reducing their uptake in fishes (Part and Lock, 1983). Accumulation of metals further depend upon the environmental conditions and their availability in water and the sediments. Differences in bioaccumulation of metals in body tissues also depend upon the animals' physiological activities. *O. nilotica* is said to have increased accumulation of copper in liver, than in *L. rohita* or *C. catla*. The accumulation of metals by various intracellular compartments is further a function of relative rate of metal binding and release, as determined by their relative metal binding affinity and competition among different metals for the binding sites *i.e.* binding to metallothionein (MT). Low levels of cadmium are known to induce MT formation but with chronic exposures cadmium binding capacity of MT decreases. Catla is, however, said to possess a strong tendency to retain lead in the kidney. Organic mercury shows strongest affinity to bind with MT, so much so that it displaces copper and zinc from the molecule. Methyl mercury however, do not bind with MT. Eaton (1985) reported binding affinity of metal to MT in the order: Hg > Cd > Cu > Zn.

Cadmium toxicity depends upon intake of copper as copper displays cadmium from MT molecules resulting in cadmium elimination. On the other hand lead is synergistic to copper uptake and elimination. There is also report for their vice-versa effects. Zinc uptake is inhibited by low level of cadmium but zinc seems to have no effect on uptake of cadmium. These interactions suggest further complicate role of essential and non-essential metals.

The Tilapia fishes (*O. mossambicus*) generally accumulate higher amount of metal residue as compared to other fishes and shows higher susceptibility to metal toxicity. The metal in combination has been found to exhibit higher toxicity effect. It has been reported that zinc and cadmium alone cause 32 per cent and 42 per cent reduction in growth, respectively. Whereas these together at sublethal concentrations exert 65 per cent reduction. Likewise cadmium with copper cause 86 per cent reduction in growth, whereas copper alone is 47 per cent toxic. Even copper and zinc together could yield 73 per cent reduction in growth in tilapia fishes. Increase in lead concentration also cause higher mortality. In common carps mortality or retarded growth occur at cadmium concentration ranging from 4-12 µg/l.

Bioconcentration factors have also evidenced greater bioaccumulation of metals by molluses than macrophytes in wastewater fed wetlands (Bhattacharya and Chakraborty, 2002).

Physiological Impact of Heavy Metals

Apart from acute toxicity, the heavy metals in water have proved dangerous and harmful because of their bioaccumulation and their impact on tissue degeneration. Beside these chemicals are also established to induce reproductive functions, thus influencing growth, survival and reproductive potential of animals/fishes. Loss of immunity, necrosis, inflammation, oedema are many other pathological conditions.

Chromic effect of copper adversely affected the growth and reproduction in fish. Precipitation of mucus on the gills of fishes exposed to copper leads to respiratory troubles. Fishes exposed to mercury exhibit neurological damage and histopathological changes in the visceral organs and gills. All metals are strongly chaleophilic (affinity for sulphur) in their mineralogy and chemistry and their affinity for sulphur is greater than oxygen. Hence they attack the sulphur bonds present in the enzymes and thus inhibit various metabolic activities that are taking place in the organism. They also exert toxic effect by combining with one or more reactive groups essential for normal physiological function.

Most important are quantitative differences between the chemicals *i.e.* chromium is not nearly as harmful to gill tissue as is nickel, when fish are immersed in solution of comparable toxicity (Hughes *et al.*, 1979). Zinc at a dose level of 71.5 µg/l causes death due to hypoxia because of rapid drop of blood oxygen and pH. Cadmium causes both anaemia in a variety of fish species at low and high concentration as well as abnormally large number of malformed RBC and altering the leukocyte differential count. Cadmium is further reported to cause inhibition of serum antibody titres in cunners, but a six fold stimulation in stripped bass (Robohm, 1996). Newman and Maclean (1974) has also reported a dose dependent 3-4 fold increase in neutrophyls and a dose dependent decrease in lymphocytes in the cunners (Marine teleost fishes). A study conducted by Verma (1999) from India has also evidenced significant increased in Hb content and RBC count in case of Indian major carps, exposed to copper and zinc while WBC counts were decreased. But in the group exposed to cadmium a significant decreae in Hb content, RBC count alongwith WBC count, as in the case of other heavy metals was observed. In case of copper and zinc the lymphocytes were decreased significantly, while the granulocyte and monocyte population tend to increase and with cadmium monocyte and lymphocytes decreased significantly. A significant decrease in the antibody titers was observed in the group exposed to copper and zinc. The serum protein levels were also decreased in all the cases of heavy metals. Likewise, chronic exposure of the cyprinid *Barbus* to lead for upto 60 days at 47 µg/l also resulted in severe reduction (12 to 31 per cent) in RBC count, haematocrite, haemoglobin concentration and MCV (Tewari *et al.*, 1987).

Other major physiological aspects of metal toxicity are alterations in cellular enzyme activity. For example lead is said to act as a stimulator of alkaline phosphatase, but a strong inhibition of both acid phosphatase and xanthine oxidase. Cadmium along with mercury result in generally enzyme inhibition at high concentration. There was a clear cut increase in carbonic and hydrase activity in flounders exposed to cadmium (Gould, 1977). Shaffi (1993) also reported significant inhibition in certain metabolic enzymes such as lactate dehydrogenase, succinic dehydrogenase and malate dehydrogenase in the freshwater fish species, following their 24 or 48 h exposure to mercury.

Prolonged cadmium exposure of male brook trout is reported to cause their hormone levels to continue to rise after spermiation at which time they normally decline and the testis further exhibit histopathological damage from cadmium possibly due to inhibition of the system responsible for normal metabolic breakdown of steroids. Hg both organic and inorganic forms also cause histological lesions in the Leydig's cells of the freshwater catfish *Clarius* (Kirubagaran and Joy, 1992) and lead at low concentration is also said to cause marked suppression of spermatogenesis in trouts. The metals further interfere with the dynamic cell surface measurements due to complexion of ions particularly the divalent cations like cadmium.

There are many other reports on the physiological and biochemical changes in animal tissue on account of heavy metal toxicity. The biochemical components like proteins, lipids, cholesterol and phospholipids in the liver of *Cyprinus carpio* were found significantly reduced following 60 days exposer to nickel (7.5 to 15.0 mg/l) and chromium (10.0 to 20.0 mg/l).

Heavy Metal Management

Today fishes have become the major diet and there have been attempts to devise ways to enhance fish production. The need of hour is to rationalize the use of industrial effluents on one hand and to minimize their interference with other biotic components including the fishes. Wastewater management technology has received considerable attention in recent years to meet the stringent regulation of water quality. Research in this field has resulted in development of various treatment techniques involving various chemicals and biological materials like bacteria, algae for the precipitation or elimination of various harmful toxicant. Organic loading in terms of biochemical oxygen demand (BOD) is found to be the limiting factor for ponds and rivers. Commonly around 9000 million cubic meters of waste water is discharged in India, of which nearly 32 per cent is currently treated with primary treatment systems. In eastern part of Calcutta alone a number of sewage fed ponds covering about 25000 hectare area are established which approximately produce about 7000 tonnes of fishes annually (Gupta and Mitra, 1997).

Among various biological materials in use under wastewater management technology are also certain plant types being used for reducing heavy metal toxicity levels in water such as neem bark is used for the elimination of mercury, pine leaves for nickel, cypress leaves for lead and Ipomea leaves for cadmium. *Chlorele vulgaris* is reported to remove these metals upto 100 mg/l of the sewage waste (Shrivastava and Sohani, 2002).

Bioaccumulation method is effectively used for removal of toxic metals in wastewater treatment and industrial waste. This process is a component of commercial water treatment circuit and uses a combination of flocculation and settling as a mechanism by which metals are removed. Micororganisms and their extracellular products of metablism play an active role in flocculation. Majority, the metals present in the system seem to be removed from solution through adsorption onto the floes or onto the sludge by physicochemical interactions involving the ion-exchange types of mechanism (Dahiya, 1990).

For absorption and adsorption of metals by microorganisms involves more than one physiological activities. These mainly include- metal transport mechanism, which accumulate and store selective metals including Mg, Ca, Mn, Fe, Zn, Si, K, Cu, Co, Mo. The transport of metal in microbial cells has been studied in *Chlorella pyrenoidosa*, *Dunaliela tertiolecia*, Baker's yeast, *Neocosmospora vasinfectum* and *Escherichia coli* (Kelly *et al.*, 1979).

Failla *et al.* (1976) reported the accumulation and storage of Zn^{2+} by *Candida utilis*. Metal accumulation affected by chemical factors such as cations soluble amino acids and physical factors including temperature, conductance, pH etc. Mostly, the binding efficiency shows in acidic pH. When pH goes to alkaline the binding efficiency decrease (Gloaguen and Morvan, 1997). In general, binding ability of the barks for the metal is, Pb>Cu>Cr Ni>Fe.

Krishnan *et al.* (2002) have further stated that heavy metals such as mercury, lead, cadmium and cobalt from wastewater can be effectively sorbed by sulphurized activated carbon. Sulphurized activated carbon was prepared from sugar industry waste, baggase pith, by single-step steam pyrolysis in the presence of SO_2 and H_2S at 400°C. The maximum removal took place at the pH range of 4.0-8.0. Maximum removal of 98.2 per cent for Pb(II), 97.5, per cent for Hg (II), 91.5 per cent for Cd(II) and 73.6 per cent for Co(II) was observed at an initial concentration of 100.0 mg/L at 30°C and pH 6.0.

Chapter 16

Fish Production in Relation to Aquacultural Development in India: A Review

The aquaculture production has been steadily increasing over the last decade. The global production of farmed fish and shellfish (according to FAO, 1995 categorization) increased from 10.64 million metric tones (mt) in 1987 to 26.38 million metric tones in 1996, indicating an increase of 148 per cent. The corresponding increase in value was 110 per cent, from 16.38 billion to US$ 41.55 billion. The boom in aquaculture production is reflected in the production figures for India as well, which shows an increase from 0.78 million mt in 1987 to 1.77 million mt in 1996 (126 per cent), and the corresponding value increase was from 0.83 billion to US$ 1.98 billion (139 per cent). This estimate by FAO, might appear conservative when, according to the 'Marine Products Export Development Authority of India' (MPEDA), the export earnings from shrimp alone, exceeded one billion dollars consecutively for the last four years. Asian countries claim the lion's share (91 per cent) of global aquaculture production, adopting a spectrum of different culture systems and intensities, from the traditional to the most modern, with other continents trailing behind, some excelling however in production of certain categories of fishes *i.e.* diadromous fishes, such as salmonids. The aquaculture boom and increased socio-economic benefits, such as increase in employment and social well being, concomitant with increase in the extent and intensity of aquaculture, are alleged to have created several problems, particularly those a deteriorating environment owing to worsening water and soil quality, as well as owing to incursions by aquaculture into the other sectoral activities, such as agriculture, capture fisheries, forestry, public health, tourism, urbanization and housing. The impact can be two-fold, one affecting aquaculture itself by auto pollution and consequent outbreak of diseases, leading to crop failures and collapses eventually. The first such national complete collapse in shrimp culture was observed in Taiwan, followed by partial collapses of cultured shrimp production in China, Indonesia, Thailand, Philippines as well as in India in 1995–97 (FAO, 1999). This has resulted in several complex social problems and litigation at the national level in India, still awaiting legal settlements to decide on the directions Indian Coastal Aquaculture will take. Sustainability of aquaculture itself has been questioned and several international, regional and national level studies and discussions on these subject have taken place recently (Kutty, 1999).

Fresh water aquaculture is no longer in its infancy in India. There is a growing awakening on the use of quality seed, good nutritional feed, and maintenance of suitable water quality and a healthy, disease-free environment. More advances were made in the 1970s with All India trials and area-technology of fresh water aquaculture, the Indian Council of Agricultural Research (ICAR) in 1980, established a Freshwater Aquaculture Research Training Centre (FARTC), with modern research laboratories for upgrading technical competence and initiating in-depth studies for optimizing productivity by a better understanding of the nutrient dynamics in tropical pond environment nutrition and diseases of cultivated species and genetic improvement for producing strains of carps with better cultural traits. This was another turning point in freshwater aquaculture research in India, which gave disciplinal research thrust in this area.

To provide a strong base for development of aquaculture and fisheries in India, several research and training promotion and regulatory institutions have been set up over the years. Two fisheries research institutes, *i.e.* Central Marine Fisheries Research Institute (CMFRI), Cochin, and Central Island Fisheries Research Institute, (CIFRI), Calcutta, were transferred to the Indian Council of Agricultural Research (ICAR) in October 1967 and have since been operating continuously. In 1985, CIFRI was divided into three institutions:

1. Central Inland Capture Fisheries Research Institute (CICFRI), Barrack pore, Kolkata
2. Central Institute of Freshwater Aquaculture (CIFA), Kausalyaganga, Orissa, and
3. Central Institute of Brackish water Aquaculture (CIBA), Chennai, to which were also transferred some activities on brackish water culture from CMFRI. In addition there are the Central Institute of Fisheries Technology (CIFT) at Cochin, which conducts research in handling and processing of fish and the Central Institute of Fisheries Education (CEFE) at Mumbai which conduct the training and the educational programmes. Two additional centers, namely, National Bureau of Fish Genetic Resources (NBFGR) and National Research Centre on Coldwater Fisheries (NRCCF) have also been established.

The Ministry of Agriculture (GOI) has established the Central Fisheries Nautical and Engineering Training (CIFNET) and the Central Institute of Coastal Engineering for Fisheries (CICEF), CIFNET is the major institution providing trained manpower for operating larger vessels and CICEF provides aquaculture engineering support to brackish water development project. Besides the Integrated Fisheries Project (IFP) involved in processing and marketing of fish is also under the Ministry of Agriculture (GOI).

There are two major promotion and regulation institutions under the Ministry of Commerce. These are the Marine Products Export Development Authority (MPEDA) and Export Inspection Agency (EIA). The newly formed Ministry of Food Processing Industries has now taken over the Fishery Survey of India (FSI), involved in exploration of marine fishery resources, and is now in the process of forming a national fisheries advisory board. In additional, many states have their own training centers for training of fishermen, extension workers, fisherwomen and farmers. This section also provides a background of most of these important national institutions.

In the economy of India, Fisheries play an important role. It helps in augmenting food supply, generating employment, raising nutritional level and earning foreign exchange. The Department of Agriculture and Cooperation has been undertaking directly and through state governments/UTs/various production, input supply and infrastructure development programmes and welfare-oriented schemes besides formulating appropriate policies to increase production and productivity in the fisheries sector (RRTD, 1998).

The freshwater sub-system of Indian Fisheries involves three types of activities:

1. Capture fisheries, practiced in reservoirs, which in real way is stocking-cum-capture fisheries, practiced in, most of the reservoirs in India
2. *Capture Fisheries,* mainly undertaken in rivers
3. *Culture Fisheries,* usually undertaken in ponds and tanks.

India has a 29000 km stretch of rivers and canals, 1.45 million ha reservoirs and 0.75 million ha tanks and ponds. There has been a steady growth in fish-production in freshwaters from inland fisheries as shown in the Table 16.1 (Srivastava *et al.,* 1991).

Table 16.1: Freshwater Fish Production from Inland Fisheries (1950–1997)

Year	*Production (in lakh tonnes)*
1950–51	2.18
1960–61	2.80
1970–71	6.70
1980–81	8.87
1981–82	9.99
1982–83	9.40
1983–84	9.87
1984–85	11.03
1985–86	11.60
1986–87	12.29
1987–88	13.00
1988–89	13.80
1989–90	14.60

Fish production has increased at an average rate of 4.4 per cent per year during the five years of the Eight five year Plan as shown in the Table 16.2.

Table 16.2: Fish Production During Eight Five Year Plan (1992–1997) (in lakh tonnes)

Year	*Marine*	*Inland*	*Total*
1992–93	25.76	17.89	43.65
1993–94	26.49	19.95	46.44
1994–95	26.92	20.97	47.89
1995–96	27.07	22.42	49.49
1996–97 (Target)	28.57	22.83	51.40
1997–2000	26.90	–	–
2000–2001	23.30	–	–

The state wise statistics of Inland fresh water fish landing from 1967 onwards have also shown a continuous increase in inland fish production. However, despite vast resources, fish production. Potential of fresh waters in India, especially in ponds and tanks, rivers, lakes and reservoirs etc, has not been fully utilized (Saktyhivel, 1999).

In the system of Inland freshwater capture fisheries, species wise production of farmed teleosts and crustaceans in India, based on FAO statistics for 19 per cent, and percentage increase in production of the different species are given in Table 16.3. Freshwater fish farming, mainly through composite culture of Indian major carps and Chinese carps, account for most of the production.

Inland Capture Fishery Resources

Inland capture fishery resources of India are well known for their variety and production potential comprising 45,000 km of rivers, 1,26,334 km of canals, 2 million hectares of estuaries and 2.05 million hectares achieved from all these resources in much below it potential.

The extensive network of Indian rivers and canals constitute one of the major inland fisheries resources the country besides being the source of the origin germplasm. The present day riverine fishery is below the subsistence level with average yield of 0.3 tonne per km which is only about 15 per cent of their actual potential. Reservoirs from the most important inland open water fishery resource in the country with an average of 2.05 million hectares. However, at the present level management and utilization, the average field yield only 20 kg/hectares/yr. The three pronged strategy reservoir fisheries development comprising enlargement of mesh size, balanced fishing effort and sustain stocking support has been paying rich dividends and production of 50–100 kg/hectares/yr can be realized from large (> 5000 ha) and medium (> 1000 to >5000 ha) reservoirs.

Table 16.3: Species-wise Aquaculture Production (mt) in India (1987–1996)

Species-wise *1*	*Production (mt) in 1996* *2*	*% Increase 1987–1996* *3*
A. teleosts		
Catla catla	417.650 (23.6)	(131)
Cirrhinus mrigala	409.980 (23.2)	(233)
Labeo rohita	422.260 (23.9)	(131)
Cyprinus carpio	13.705 (0.8)	(69)
Ctenopharyngodon idellus	15.305 (0.9)	(15.9)
Hypophthalmicthys molitrix	3.755 (0.2)	(1.50)
Anabas testudineus	55.000 (3.1)	(51)
Clarias spp.	57.638 (3.3)	(56)
Other (*Osteicthyes*)	285.602 (16.2)	(34)
B. Crustaceans		
Penaeus monodon	87.187 (4.9)	(481)
P. indicus	162.00 (0.01)	(–)
Macrobrachium spp.	178.00 (0.01)	(–61)
Total	**1.768.422**	**(125)**

Figures in brackets denote percentage composition (col. 2).

Percentage increase (col. 3) in production during 1987–1996, expect for *Ctenopharnygodon idellus, C. carpio,* and *H. molitrix* (1993–1996) and *M. rosebergii* (1992–1996).

The extensive floodplain wet-lands in the form of oxbow lakes (manus, chaurs, jheels, bheels–as they are called locally), especially in the states of Assam, West Bengal, Bihar and Eastern Uttar Pradesh, occupy an important position in the inland fisheries resources of India because of their magnitude as well as their production potential. These water bodies are rich in nutrients, as reflected by rich organic carbon and high levels of nitrogen and phosphorus in their soil. The yields of floodplain wetlands can be increased to average one tonne of fish per hectare from the present average yield of 200 kg/ha/yr by following appropriate culture practices. The practice of pen culture is an ideal operation for these resources. The other approaches for sustainable fish production from inland capture fisheries sector include cage culture, ranching of fish seed for higher production, provision of fish passage for migratory species and endemic genome conservation through fish sanctuaries.

Table 16.4: Inland Fishery Resources

Resource	*Area (hectares)*
Ponds and tanks	2.36 million ha
Oxbow lakes and derelict water bodies	1.3 million ha
Reservoirs	2.05 million ha
Rivers	45,000 million ha
Canals	1,26,334 million ha
Estuaries	2 million ha
Brack	1.42 million ha

Development of Sustainable Aquaculture/Fisheries

Aquaculture can be defined as tending the confined waters for growing aquatic organisms and harvesting the production for human benefits. One of the main objectives of the aquaculture technologies was to increase productivity of the cultured waters, *i.e.* increasing the intensity of culture causes changes in the ambient water and soil, leading to deterioration of the environment. With suitable inputs, which would raise the costs, the water and soil conditions can be recorded, though not easily to the original purity, but to tolerable levels. If the aquaculture units are small or dispersed, the problem of greater environmental deterioration may not be much and can be managed. However, if the density of the units were too high for the environment, *i.e.* the water shed or the ecosystem, *i.e.* beyond its carrying capacity, then serious problems could arise.

When there is a high price for the aquaculture product, such as the shrimp, unless proper plans, policies and enforcement through legislation are in place, haphazard development of aquaculture ventures could take place. This would be unsustainable, creating impacts not only to aquaculture itself but also to other competing/

co-operating sectors. This would cause problems concerning deterioration of environment, and inroads to other sectors through demands for water, salination, restriction of movements of people, navigation and fish capture, urbanization, tourism, etc. Above all, this would affect the sustainability of the system itself through auto pollution, which results in diseases and mortality of the cultured organisms to the extent of making the system itself uneconomical due to poor returns or its complete collapse, as already pointed out. In many of these cases uncontrolled increase in the number of shrimp culture units/ponds per unit are/castline has gone much beyond the *Carrying capacity* of the ecosystem (environment) and has led to the collapse of the systems (Kutty, 1999).

Augmenting Marine Fish Production

1. To upgrade technologies and capabilities in the artisanal and small mechanized sector to increase productivity through diversification.
2. To establish an oceanic fishery for tunas, squids and sharks.
3. To improve the coastal productivity through artificial reefs, FADs and sea-ranching.
4. To exploit the Perch resources from untrawlable grounds including island territories through targeted fishing.
5. To harvest the deep sea shrimps and lobsters.
6. To reduce by catch and prevent discards.
7. To stop destructive fishing methods and capture of juveniles and subadults.
8. To reduce post-harvest losses.
9. To popularize PFZ advisors among fisher-folk.
10. To develop and adopt appropriate eco-friendly mariculture technologies and increase fish production through mariculture.
11. To launch scheme of motorization of traditional craft by liberal financing through various financial schemes.

Options for Sustainability

The issues pertaining to marine fisheries in India are not unique to the country but common to most of the tropical developing countries.

Under the present scenario, there is need for policy support on:

1. Effective implementation of the regulatory measures for optimizing the exploitation of the fishery resources in the EEZ and beyond.
2. Exploitation of the deep-sea and oceanic resources in the EEZ and beyond.
3. Effective prevention of third country fishing in the Indian EEZ.
4. Increasing the pace of motorization artisanal craft to improve their capabilities.
5. Monitoring and regulating fishing effort in the inshore waters.
6. Improvements in the system of data generation as well as quality of data on exploited stocks.
7. Protection, conservation and sustainable utilization of marine biodiversity.
8. Improvement in domestic and export marketing through value addition.
9. Manpower development in R and D, fishing and processing sectors.
10. Development of Marine Fishers Information System.
11. Development of infrastructure–fishing harbours–berthing facilities, capacity utilization of post harvest infrastructure.
12. Stock enhancement through artificial reef habitats, fish aggregating devices and sea ranching.
13. Development of and adhering to a National Code of Conduct for Responsible Fisheries.

In summing up, India could benefit considerably from global experience in aquaculture development. India has achieved considerable production increases in aquaculture, especially in the production of fresh water fishes and shrimps. While progress in research and development of new technologies have already been made, for example in mollusc culture and seaweed culture, these have not yet taken off as serious commercial ventures, and the production gaps between India and China/other Asian countries are very wide (million(s) of tonnes). We could utilize much more effectively the diversity of fresh and marine living resources for aquaculture. Along with the development of technologies, there must be greater efforts to spread the message at the grass roots level, through a newly oriented extension set up. India could also benefit much by collaborating with other countries in Asia, which have developed and perfected some of these inland and coastal culture systems, again enabling technology transfer at the grass roots level.

Strategies for Future

The fisheries sector contributes about Rs. 22,200 crores to the GDP, which is about 1.4 per cent of the total GDP and 4 per cent of the agriculture sector. The following are the main objectives envisaged during the X plan:

1. To contribute to food and nutritional security.
2. To increase contribution to GDP.
3. To augment export earnings.
4. To generate employment.
5. To ensure sustainable development of the sector.

References

FAO, 1995. *FAO Technical Guidelines for Responsible Fisheries*, No. 5, FAO, Rome, 1995, p. 40

FAO, 1999. *Review of the State of World Fishery Resources: Inland Fisheries*, Circular No. 942, pp. 53, FAO, Rome, Italy, 1999.

Kutty, M.N. 1999. Aquaculture Development in India from a global perspective. *Curr. Sci.,* 76(3): 333–341.

RRTD, 1998. India 1998: A reference annual ed. Research reference and Training Division (RRTD, 1998). Publ. Ministry of Information and Broadcasting, Govt. of India, 1998.

Sakthivel, M. 1999. Strategy of increase export of fish and fishery products from India. *Curr. Sci.,* 76(3): 405–412.

Srivastava, W.K., B.M. Dholakia, S. Vathsala and K. Chidambaram (Eds.) 1991. *Fishery Sector of India*. Oxford and IBH Publishing Co., New Delhi, pp. 437.

Chapter 17

Data Requirement for Planning and Management of Inland Fisheries

Planning and management for sustainable exploitation of aquatic resources, both inland and marine sector depends on base line reliable data available. They become all the more important in this highly dynamic situation of modern time. Many development projects were started and completed with little access to scientific information of fish and fisheries resources and also of socio-economic information of those who are undertaking fisheries. This whole situation becomes more complex in the country. Under above circumstances the country needs a most pragmatic, ecologically sound, economically feasible and socially acceptable fisheries development plan. For proper fisheries research, planning and development and also to monitor progress of project, quick and highly reliable, effective geo-referenced data are essential. Fish is an aquatic vertebrate so it gets all requirement of life-oxygen, food, shelter etc. from water. Similarly, the parasites, the pathogens and many of the predators etc. of fishes are also aquatic. Qualitative and quantitative changes in aquatic environment are bound to affect this habitat which in turn affect life processes of fishes, therefore by manipulating the factors in aquatic environment it is possible to bring about qualitative and quantitative changes or improvements in the fisheries resources. Thus there should be all information about the geographical region with regards to physical, chemical and biological factors. Information may be collected through conventional method along with Geographical Information System (GIS) and remote sensing at regular intervals on all physical, chemical and biological parameters relevant to planning and development of inland fisheries of the geographical area concerned and process the data to a useable form.

Fisheries is a complex subject and hence it is difficult to come at perfect parameter and to make a list. Even then a sequence of factors has been made in which information for fisheries development is present. Mode of data bank is suggested for different technologies.

Ponds

Physical Factors

Number of ponds-perennial, seasonal

Size (area) of individual ponds-classification according to size.

Dimensions of individual ponds

Configuration of bank (shore-line)

Gradient (of slope).

Soil type

Nature and extent of erosion, if any

Vegetation on banks-north, east, south, west

Depth of water at different times of the year contour map

Outlets/inlets if any

Turbidity

Temperature

- At surface
- At 2 m depth
- At bottom

Available transport facilities

Railway tracks, highways in the vicinity

Chemical Factors–Seasonal Variations

Dissolved oxygen

Dissolved carbondioxide

Other gases

Dissolved/suspended organic matters-kinds, quantity

Dissolved / suspended inorganic matters-kinds, quantity

pH

Biological Factors–Seasonal Variations

- Vegetation types in pond water
  - Emergent
  - Submergent
  - Floating
- Phytoplankton
  - Types
  - Number of different types per unit volume of water
- Zooplankton
  - Types
  - Number of different types per unit volume of water
  - Diurnal migration pattern
- Benthos
  - Types
  - Number of different types per unit volume of bottom soil
- Species composition of standing fish population (relative abundance of various species)
  - Cultured fishes
  - Weed fishes-predatory fishes
- Crustacean population
  - Species composition
- Molluscan population
  - Species composition

Other Factors

- Ownership pattern
- Management types
  - Private management
    - Single owner management
    - Multiple owner management

Government management

NGO management

Combination

Management problems, if any

Fish culture practices, if any

Pre-stocking treatments, if any

Stocking pattern

Species combination

Sources of fry/fingerlings

Uses other than fish culture

Irrigation

Washing

Bathing

Cattle bathing

Duck raising

Factors Relevant to Fisheries Practices

Inland Capture Fisheries

Life history of fishes

Commercial fishes

Weed fishes

Aquarium fishes

Others

Life history of non-piscine fisheries organisms

Pollution status-pollutants

Sediment transport

Fish seed sources (fin fishes, shell fishes)

Natural sources

Hatcheries

Hilsa fishery

Factors for migration

Responses to various stimuli at different stages of life

Rheatactic responses

Chemotactic responses

Effect of barrages

Effect of other human interferences

Nets across river mouths

Pollutants

Destruction of juveniles

Pangasius fishery

Masheer fishery

Impact of open water stocking

Impact of new fisheries management policy

Reservoir fishery reaction to transition from riverine

Fish transport-fry/fingerling transport

Fish marketing

Estuarine fishery-management under the Forestry Department versus management under fishery department

Conservation and management-impact of participatory management practices.

Inland Culture Fisheries

Life history of culturable fish species (breeding habits, induced breeding)

Food and feeding habits

Soil water interaction-fertilization

Effect of various fertilizers

Hatchery operations and management

Common problems

Shrimp hatcheries (Penaeus, Macrobrachium)

Fish seed industry

Fish cum duck farming

Predatory and weed fishes (control measures)

Aquarium fishes (potential for export market)

Aquatic insects-control measures

Fish food organisms (commercial production)

Feed formulation for various stages of life history of various culturable species of fish, supplementary feeding

Stocking rates

Composite culture

Aquatic weed control

- Mechanical control
- Chemical control
- Biological control

Transport of brood fishes

Transport of fish seed

Causes of mortality in transport operations

Use and effect of various drugs

- Fish toxicants-anaesthetics
- Antiseptics
- Antibiotics

Culture of exotic fishes

Culture of carnivorous fishes

Fish pathology, fish parasitology

Brackish water ponds

- Water quality (salinity)
- Soil quality
- Collection of seed from natural sources

Coastal aquaculture

Remote sensing and fisheries development potentials.

These informations would help in proper aquatic resource utilization, management and conservation of fisheries resources while preserving the environment and also in training manpower at primary, secondary and tertiary levels etc. for environment monitoring and management.

Chapter 18

Causes of Fishes Decline and Efforts for Revival and Statistics of Major Inland Fisheries

Statistics of Major Inland Fishery

The total annual catch of the marine and inland fish of the world rose from 19.6 million tones in 1948 to 76.6 million t in 1983. The estimated world fish production during 1986 is 90 million t showing a rise of 4.7 per cent over the estimated catch of 1985. For the purpose of recording fish catch statistics on global basis, the world has been divided into 26 international major fishing areas. These consists of 7 inland fishing areas (each comprising one of the seven continents) and 19 major marine areas covering the 3 oceans, the Atlantic, the Indian and the Pacific ocean, each with their adjacent seas.

Potentially the vast and varied inland fishery resources of India are one of the richest in the world. They pertain to two types of waters, namely the fresh and the brackish. The former includes country's great river systems, an extensive network of irrigation canals, reservoirs, lakes, beels, tanks, ponds etc. and the latter, the sprawling estuaries behind the mouths of countries rivers, a large no. of lagoons, brackishwater impoundments and vast areas of mangrove swamps containing tidal waters of fluctuating salinity.

In India the total fish production stood at 3677 thousand metric tones in 1999–90 it showed a steady increase upto 5388 thousand metric tones in 1997–98 then it decrease and increase, it being 5262 thousand metric tones in 1998–99 and 5657 thousand metric tones in 1999–2000 and goes on increasing.

Estimation of Inland Fish Catches

The question of compilation of fishery statistics has been receiving the attention of both the central and state government as well as other organizations of a long time. The statistics of total fish catch from inland resources are not available beyond the rough estimates provided by the "Reports on the Marketing of Fish in the Indian Union" (1948, 1951 and 1961) and estimates of annual marketable surplus furnished by the states. The collection of fish production statistics from inland water is beset with serious difficulties, the main in respect of riverine and estuarine fisheries are:

1. Highly dispersed and isolated nature of fishing and landing areas.
2. The diversity of fishing gear and tackle employed and a high percentage of subsistence fishing.
3. The innumerable landing places.
4. The migration of fisherman from one area to the other for fishing.
5. The prevailing system of fish merchants buying off catches from the fishing boats at the fishing spots.
6. The multispecies composition of catches and the landing of catches in unsorted conditions.
7. The isolated nature of fishing area in lower estuaries where transport for making direct observations are very limited (Pillay, 1960).

Major Technologies Developed for Inland Fisheries

The Central Inland Capture Fisheries Research Institute (CIFRI) is the oldest premier research institute in the field of inland fisheries research and training in India. Establisehd initially as Central Inland Fisheries Research Station by the Government of India at Kolkata in 1947, it assumed the status of an Institute and shifted to Barrackpore (North 24-Parganas, West Bengal) in 1959. The institute came under the administrative control of Indian Council of Agricultural Research in October 1967.

Table 18.1: Inland Fish Production of Different States in India ('000 tonnes)

State/Union Territory	*1994–95*	*1995–96*	*1996–97*	*1997–98*	*1998–99*	*1999–2000*
Andhra Pradesh	195.13	203.97	207.31	226.31	260.83	380.58
Arunachal Pradesh	1.8	1.85	1.85	2.13	2.3	2.4
Assam	153	155.06	154.61	155.13	155.71	159.77
Bihar	195.37	239.58	249.78	208.54	202.29	254.74
Goa	3.44	3.61	3.3	3.24	3.18	3.51
Gujarat	70.1	60	65.28	70.8	80	70.33
Haryana	24.13	28.01	30	32.05	32.52	30
Himachal Pradesh	5.29	5.94	6.27	6.69	6.79	7
J&K	16.1	16.52	17.8	18.53	18.85	19.01
Karnataka	70.29	87.35	101.65	95.28	95	126.65
Kerala	48.19	49.59	52.1	57.51	65.86	73.9
Madhya Pradesh	80.18	91.28	110.53	115.16	119.59	127.43
Maharashtra	89.88	77	109	127	125.5	135.39
Manipur	12.01	12.5	12.71	13.7	15.31	15.51
Meghalaya	3.95	3.58	3.58	3.09	4.53	4.68
Mizoram	2	2.5	2.55	2.7	2.78	2.89
Nagaland	2.5	3	4	3.7	4.5	5
Orissa	134.77	134.85	143.5	153.43	159.9	135.3
Punjab	24	26	32	36	44.5	47.18
Rajasthan	13.97	12.4	14.3	15.1	12	12.97
Sikkim	0.1	0.15	0.15	0.14	0.14	0.14
Tamil Nadu	108	108	109	109.5	110.2	112
Tripura	25.1	25.71	27.47	27.91	28.41	29.34
U.P.	139.9	145.7	149.43	160.02	183.03	192.71
West Bengal	669.22	740	765	786.02	823.5	865.7
Andaman & Nicobar Islands (P)	0.05	0.05	004	0.04	0.04	0.06
Chandigarh	0.08	0.08	0.08	0	0	0
Dadar & Nagar Haveli	0.08	0.04	4	0.02	0.02	0.03
Daman & Diu (P)	0.04	–	–	–	–	–
Delhi	3.9	4	4.1	4.2	4.42	4.3
Lakshdweep (P)	–	–	–	–	–	–
Pondicherry	4.19	4	0.04	4.1	4.11	4.21
Total	**2096.76**	**2242.32**	**2381.43**	**2438.04**	**2565.81**	**2822.73**

Sources: State Governments/Union Territory Administrations.

Table 18.2: State-wise Fish Production in India for 1995–2000

State/Union Territory	*1995–96*	*1996–97*	*1997–98*	*1998–99*	*1999–00*
Andhra Pradesh	355.96	359.36	372.86	410.82	547.06
Arunachal Pradesh	1.85	1.85	2.13	2.30	2.40
Assam	155.06	154.60	155.13	155.71	159.77
Bihar	239.58	249.78	208.54	202.29	254.74
Goa	87.82	97.06	92.05	69.02	650.62
Gujarat	660.00	725.35	816.50	630.00	741.28
Haryana	28.01	30.00	32.05	32.52	30.00
Himachal Pradesh	5.94	6.27	6.69	6.790	7.00
J&K	16.52	17.8	18.53	18.55	19.01
Karnataka	304.87	324.43	285.13	255.61	292.30
Kerala	582.14	631.03	583.86	649.22	694.40
Madhya Pradesh	91.28	110.53	115.16	119.59	127.43
Maharashtra	464.00	590.00	580.00	520.38	533.29
Manipur	12.50	12.70	13.70	15.31	15.51
Meghalaya	3.58	3.58	3.09	4.53	4.68
Mizoram	2.50	2.55	2.70	2.78	2.89
Nagaland	3.00	4.00	3.70	4.50	5.00
Orissa	258.04	276.96	309.51	284.23	261.24
Punjab	26.00	32.00	36.00	44.50	47.18
Rajasthan	12.40	14.30	15.10	12.00	12.97
Sikkim	0.15	0.15	0.14	0.14	0.14
Tamil Nadu	448.00	459.79	464.60	469.75	475.00
Tripura	25.71	27.47	27.91	28.41	29.34
U.P.	145.70	149.43	160.02	183.03	192.71
West Bengal	893.00	937.00	950.02	995.00	1045.70
Andaman & Nicobar Island (P)	25.73	26.44	27.27	27.44	28.20
Chandigarh	0.09	0.09	0	0	0
Dader and Nagar Haveli	0.04	0.04	0.02	0.02	0.03
Daman and Diu (P)	15.28	15.28	18.81	26.85	15.95
Delhi	4.00	4.00	4.20	4.42	4.30
Lakshadweep	9.82	11.75	10.55	13.54	13.60
Pondicherry	40.82	42.65	42.52	42.70	42.83
Deep sea fishing sector	30.00	30.00	30.00	30.00	30.00
Total	**4949.39**	**5348.24**	**5388.49**	**5262.25**	**5656.55**

Source: Handbook of Fishery Statistics, 2000.

Table 18.3: State-wise Marine Fish Production in India

State/Union Territory	1990–91	1991–92	1992–93	1993–94	1994–95	1995–96	1996–97	1997–98	1998–99	1999–00
Andhra Pradesh	120.35	125.79	113.07	154.32	150.26	151.99	152.05	146.55	150.00	166.48
Goa	53.18	47.11	101.49	102.11	98.46	84.21	93.76	88.81	65.84	62.11
Gujarat	500.00	516.85	589.00	619.84	645,26	600.00	660.07	745.71	550.00	670.95
Karnataka	183.83	181.41	174.19	174.52	173.75	217.51	222.78	189.86	160.61	165.65
Kerala (P)	514.24	524.76	496.24	559.20	548.37	532.55	578.92	526.34	583.34	575.50
Maharashtra	325.00	390.86	387.55	350.40	357,00	387.00	481.00	453.00	394.88	397.90
Orissa	78.00	87.88	119.38	103.93	122.89	123.20	133.46	156.08	124.33	125.94
Tamil Nadu	288.95	301.00	308.00	317.72	330.50	340.00	650,79	355.10	359.55	363.00
West Bengal	125.00	142.00	145.00	153.00	151.20	153.00	172.00	164.00	171.50	180.00
Andaman & Nicobar Islands (P)	15.15	25.19	24.17	25.08	26.12	25.68	26.40	27.23	27.40	28.15
Daman and Diu (P)	7.73	15.94	13.43	11.53	11.50	15.28	15.28	18.81	26.85	15.95
Lakshadweep (P)	7.60	5.81	9.73	9.41	9.75	9.82	11.75	10.55	13.54	13.60
Pondicherry	30.62	32.68	35.00	37.78	36.75	36.82	38.55	38.42	38.60	38.62
Others	50.00	50.00	60.00	30.00	30.00	30.00	30.00	30.00	30.00	30.00
Total	**2299.65**	**2447.28**	**2576.25**	**2648.84**	**2691.81**	**2707.06**	**2966.81**	**2950.46**	**2696.44**	**2833.85**

Table 18.4: Fishermen Population by States in India (in hundreds) (1992)

State/Union Territory	Male	Female	Children	Total	No. Family Engaged in fishing		Family Members Engaged in Fishing			
					Full Time	Part Time	Marketing in Fish	Repair of Fishing Nets	Processing of Fish	Other Activities
Andhra Pradesh	2768	2602	3348	8718	1428	1321	1121	504	219	260
Arunachal Pradesh	4	1	–	5	–	–	420	–	–	–
Assam	1538	807	795	3140	254	201	132	113	26	–
Bihar	1800	1322	1589	4711	255	754	331	167	56	167
Goa	53	49	47	149	24	14	20	6	3	7
Gujarat	889	882	1844	3615	557	236	147	116	34	637
Haryana	49	8	13	70	4	–	–	–	–	38
Himachal Pradesh	13	8	15	36	2	3	2	1	–	–
Jammu & Kashmir	120	129	124	373	19	22	39	11	1	–
Karnataka	434	411	615	1460	134	74	114	47	16	26
Kerala	1978	1970	2381	6329	1099	275	254	135	81	426
Madhya Pradesh	890	671	1050	2611	112	566	152	103	23	15
Maharashtra	1277	1670	1022	3969	761	387	650	272	184	45
Manipur	305	186	66	557	318	239	–	–	–	–
Meghalaya	–	–	–	0	–	–	–	–	–	–
Mizoram	1	1	–	2	–	–	1	–	–	–

Contd...

Table 18.4–Contd...

State/Union Territory	Male	Female	Children	Total	No. Family Engaged in fishing		Family Members Engaged in Fishing			
					Full Time	Part Time	Marketing in Fish	Repair of Fishing Nets	Processing of Fish	Other Acti-vities
Nagaland	–	–	–	0	–	–	–	–	–	–
Orissa	1065	977	1400	3422	225	128	132	113	4	9
Punjab	–	0	–	0	–	–	–	–	–	–
Rajasthan	26	24	37	87	5	9	–	1	–	–
Sikkim	–	–	–	0	–	–	–	–	–	–
Tamil Nadu	1460	1325	2379	5164	890	145	221	237	48	69
Tripura	27	17	21	65	8	31	13	4	–	1
Uttar Pradesh	5951	5196	5146	16293	301	750	246	89	50	376
West Bengal	2829	1181	1343	5353	884	1927	613	320	95	394
Andaman & Nicobar Islands (P)	80	64	97	241	9	22	5	5	5	1
Chandigarh	3	–	–	3	3	1	3	–	–	–
Dadar and Nagar Haveli	–	–	–	0	–	–	–	–	–	–
Daman and Diu (P)	–	–	–	0	–	–	–	–	–	–
Delhi	–	–	–	0	–	–	–	–	–	–
Lakshadweep (P)	174	175	175	524	9	21	–	–	6	2
Pondicherry	127	124	135	386	83	11	31	13	24	89
India	**23861**	**19800**	**23642**	**67303**	**7384**	**7137**	**4647**	**2257**	**875**	**2562**

Table 18.5: Major Markets of Indian Marine Products

		1991	*1992*	*1993*	*1994*	*1995*	*1996*	*1997*	*1998*	*1999*	*2000*
Japan	Q	40109	38277	43563	54063	49889	64698	71948	65568	65202	71060
	V	61066	70267	120440	165758	158183	185967	226195	237456	217892	262477
USA	Q	20937	18690	25127	33004	24794	72229	45227	46963	62899	66462
	V	15114	16594	26930	48836	34904	76628	52079	60810	82802	100468
West Europe	Q	51937	61874	74822	74966	81344	0	0	0	0	0
	V	34640	41237	62098	75499	85567	0	0	0	0	0
Middle East	Q	6560	7472	7530	7974	9964	8958	13054	19565	12839	16273
	V	2458	4268	3076	4662	8639	6086	9869	18228	10169	17144
South East Asia	Q	37875	55783	77085	124126	115404	165110	224612	133347	140155	208206
	V	10832	16041	24133	56560	46585	76515	111729	82320	85506	122955
Others	Q	5512	9218	11791	7145	8129	13328	11305	12469	11929	16660
	V	3314	3737	6603	5237	5592	9507	8432	11369	11663	19301
European Union	Q	0	0	0	0	0	72229	45227	46963	62899	66462
	V	0	0	0	0	0	76628	52079	60810	82802	100468

Q: Quantity in Tonnes; V: Value Rs. in Lakh

Source: Marine Product Export Development Authority, Kochi.

Table 18.6: Export of Marine Products from India (in quantity and values)

Sl.No.	Items		1995–96	1996–97	1997–98	1998–99	1999–2000	2000–01
1.	Frozen Shrimp	Q	95724.00	105429.00	100720.00	102484.00	110275.00	111874.00
		V	2356.80	2701.80	3134.20	3344.91	3645.22	4481.51
2.	Frozen Lobster Tails	Q	1587.00	1172.00	1289.00	–	–	–
		V	51.00	43.90	47.80	–	–	–
3.	Frozen Cuttlefish/Fillets	Q	33845.00	31778.00	37258.00	34589.00	32799.00	33677.00
		V	260.90	272.40	323.40	273.31	286.22	288.99
4.	Frozen Squids	Q	45025.00	40924.00	35095.00	32254.00	34918.00	37628.00
		V	319.60	290.50	270.90	268.93	296.80	324.43
5.	Fresh/Frozen Fish	Q	100093.00	173005.00	188029.00	108556.00	–	–
		V	372.30	636.90	726.70	495.03	–	–
6.	Live Items	Q	–	2030.00	1700.00	–	–	–
		V	–	34.00	29.30	–	–	–
7.	Dried Fish	Q	7292.00	10475.00	5669.00	–	–	–
		V	40.30	47.00	33.50	–	–	–
8.	Shark Fins and Fish Maws	Q	–	–	–	–	–	–
		V	–	–	–	–	–	–
9.	Other Items	Q	12711.00	13386.00	16058.00	25051.00	165039.00	257294.00
		V	100.20	95.00	131.70	244.69	888.43	1349.00
	Total	**Q**	**296277.00**	**378199.00**	**385818.00**	**302934.00**	**343031.00**	**440473.00**
		V	**3501.10**	**4121.40**	**4697.50**	**4626.90**	**5116.67**	**6443.89**

Q: Quantity in Tonnes; V: Value Rs. in Lakh

Source: Marine Product Export Development Authority, Kochi.

The main objective of the institute till 1987 (Sixth Plan) had been to conduct investigation for proper appraisal of inland fisheries resources in the country, and to evolve suitable methods for their conservation and optimum utilization. During this period the successful implementation of four major projects on composite fish culture; Riverine fish seed prospecting; Air breathing fish culture and Ecology and fisheries management of reservoirs proved to be the turning point in inland fisheries and freshwater aquaculture in India.

The major technologies developed by the Institute till 1987 are:

1. Fish seed prospecting from rivers.
2. Fish seed transportation.
3. Induced breeding and nursery management of crops.
4. Bundh breeding of Chinese carps.
5. Composite fish culture.
6. Air breathing fish culture.
7. Aquatic weed control.
8. Fishery management of reservoirs.
9. Integrated farming system.
10. Brackishwater fish farming.

Reorganisation and the Revised Mandate

At the beginning of 7th Plan (1987), three more institutes were carved out from this institute *viz.,* Central Institute of Freshwater Aquaculture (CIFA), Central Institute of Brackishwater Aquaculture (CIBA) and National Research Centre on Cold Water Fisheries (NRCCWF). The original institute was rechristened as Central Inland Capture Fisheries Research Institute and the mandate modified, giving emphasis on capture fisheries resources of the country.

Fish Biodiversity

The database on fish biodiversity of India containing information on systematics, habitat and distribution of 2118 fin fishes has been prepared. Karyotype images and isozyme profile of fishes have also been prepared. From this database specific information on fishes as per ecosystem has been queried, major rivers as well as biodiversity rich areas like North East India and Western Ghats can be queried. The specific information on endangered, patential cultivable and ornamental fishes can be obtained. A list of threatened fishes of India comprising of 79 species under different categories of threatened status has been drawn. A digital base map of India (1 : 5 million scale) showing 330 Indian rivers and 20 lakes and reservoirs has been prepared.

Table 18.7: Species Diversity of Indian Fishes in Major Aquatic Ecosystems

Ecosystem	*Fish Species*
Cold water	154*
Warm water	433**
Brackish water	171***
Marine	1360
Total	**2118**

*: Of these, 34 fishes are common to cold and warm water.

**: Of these, 67 fishes are common to warm and brackish water.

***: Of these, 16 fishes are found only in brackish water, 73 are common to warm, brackens and marine water and another 82 are common to brackish and marine water.

For increasing fish production in a sustainable manner, improvement and expansion of capture and culture fisheries alone are not sufficient and conserving the fish biodiversity is a necessary prerequisite. Appreciating this, the Government of India approved the establishment of National Bureau of Fish Genetic Resources in December, 1983 under the Indian Council of Agricultural Research (ICAR). The institute initially started functioning at Allahabad and subsequently shifted to Lucknow in 1994. The infrastructure consisting of laboratory, administrative farm and residential complex has become operational at Canal Ring Road, Lucknow – 226 002, U.P., India since 1999.

Also sport fisheries potentials in Himalayan regions with the changing values of the society in terms of modernization and development. "Fish Sport" has become most reputed and delighted entertainment. Millions of tourists both from home and abroad visiting scenic corridors in Indian Himalayas are all used for angling in the upland streams, rivers and lakes. In reality the angling is enjoyed by the people of all ages, social and economic strata of the society and hence in present days it is considered as the best resort entertainment and local masses residing nearby for their food also. To promote the sports fisheries following suggestions are made:

1. Conservations measures for Mahseer, snow-trout and allied fisheries may be framed and employed to check the depletion of these native fishes in these Himalayan resources.
2. Facilities like approach road to the fishing sites, rest camps, anglers' hut etc. should be developed to allure maximum number of tourists and anglers.
3. To ensure the availability of game fishes in these aquatic resources seed production units and hatcheries may be established in the fishing areas.
4. Fast growing varieties of trouts should be transplanted in the upper reaches of glacial rivers.
5. Artificial propagation of Mahseer and snow-trout may be taken up for large-scale production of its seed to stock in all suitable water in the area.
6. Fish acts and regulations may be adopted strictly so that the fishery of endangered fish species like Mahseers and Schizothoracids could be augmented.
7. Extensive teaching programmes esposing various facets of fisheries in far flung localities be undertaken so that the local masses could also be benefited from the sustained fishery.
8. Angling competitions may be organized with vast publicity by news media, pamphlets, audio and video visions etc. so that maximum number of tourists/anglers could visit the area.

Causes of Fishes Decline

With India's rapidly growing population accompanied by increasing hazards of domestic and industrial pollution to the inland water of the country, scientists envisage of rapid degradation of water quality unless concrete steps are taken immediately to abate pollution. The pollution *i.e.* water pollution is the greatest source of polluting the river water and hence affecting the fishes. The rate of increase of human population of India and rapid pace of its industrialization have created problems of disposal of waste products.

Aquatic organisms are affected by pollution both directly and indirectly in various ways. Some of the effects produced as follows:

1. An increase in osmotic pressure.
2. Violent alteration the pH of the water all acids being lethal when pH is less than 4.
3. A decrease in oxygen content of water.
4. Specific toxic ingredients, which may injure the gills and other external structures cause death by intake and absorption.
5. Destruction of spawning grounds which can be serious in respects of major carps and other fishes which require special environment for breeding.
6. Destruction of biota–the fish is affected indirectly when its habitual food organisms are destroyed. The covering of the bottom of a water body by a coating of waste matter greatly reduces the food supply of the fish.
7. Mechanical injury or blocking of gills from slit or other suspended material.
8. Blocking of migration channel long distance migratory fishes during some phase of their life history may be adversely affected by highly localized pollution of the river.
9. Pathogens being carried to human beings through fish cultured in polluted waters as well as importing foul adour to fish flesh rendering it unpalatable.
10. Cumulative effect of insecticides, pesticides, etc. on fish as well as human being, *e.g.* DDT.
11. Acceleration aging of lakes and ponds-sewage pollution, even in small quantities, may change the character of an aquatic environment over a period of years. Thus, with gradual process of aging, deep, clear oligatrophic lakes may be sedimented, becoming mesotrophic then eutrophic and eventually turning into bags.

On an overall categorization, the more important types of industries whose effluents contribute to the pollution of river systems in India are pulp and paper, textile, tanneries, sugar, distilleries, shellac, hydrogenated vegetable oils, coal washing washeries, petro-chemicals and miscellaneous industries. Domestic wastes add considerably to the river pollution.

Pesticides, used extensively in modern agriculture spread and diffuse the hazards of pollution through drainage to ponds, tanks, lakes and streams all over the country, putting the interests of agriculture and aquaculture at cross purposes. While the pollution problem centers, depending upon the sites of the factories, are distributed far and wide and certain areas notably around Kanpur, Kolkata and Mumbai, where a heavy concentration of industries occurs form major pollution areas, most of the important pollution centers of the country would be covered if reviewed according to the river systems on or around which they lie.

Thrust Areas

Stock assessment and monitoring of commercially important marine species.

Development of fuel efficient fishing crafts and gears.

Prevention of post harvest losses, development of improved transport, storage and processing of fish.

Development of value-added fishery product for export.

Extraction, production and evaluation of biomolecules from marine organisms and plants for industrial and pharmaceutical applications.

Monoculture of fish and shell fish culture of ornamental fish.

Introduction of HACEP in seafood processing.

Conservation of and sustainable management of marine resources, implementation of code of conduct of responsible fishing.

Rural aquaculture and integrated fish farming, increasing availability of fish seed and feed.

Application of molecular biology in aquaculture.

Genetic mapping of important fish species through DNA fingerprinting.

Bioinformatics.

Development of vaccines for fish diseases.

Fish nutrition and fish development.

Enhance productivity of reservoir fisheries.

Brackishwater aquaculture.

Environment monitoring of aquatic ecosystems.

Development of eco-friendly and sustainable fish/shell fish farming.

Conservation of endangered fish species.

Development of coldwater fisheries.

Human resources development.

Enhancing living standards and generate employment and provide food security to fishing community.

Island saline aquaculture.

Organic farming.

Creating inventory on biodiversity in river.

Assessment of potential for developing ornamental fish trade in the north-east.

Comprehensive nation-widc H_2O resource survey and mapping.

Development of small water bodies on turnkey basis through individual and institute consultancy.

Creation of North-eastern Fisheries Training and Extension Center (NEFTEC).

NBFGR has been identified as the lead center for consolidating Aquatic Animal Pathogen Quarantine Information System (AAPQIS) of FAO/NACA/OIE.

NBFGR has developed a "National Plan for Aquatic Exotics and Quarantine" approved by Government of India, Ministry of Agriculture.

References

Abram, Mathew. *Brackishwater Aquaculture*. CIBA Special Publication No. 17, Central Institute of Brackishwater Aquaculture (Indian Council of Agricultural Research) 75, Santhone High Road, Raja Annamalaipuram, Chennai.

Ahmad, S.H. and A.K. Singh. (1987). Seasonal fluctuations of primary production and fish yield in Mangle's Tank, Patna City, Bihar. *Geobios,* 14: 62–66.

Jhingran, V.G. and P. Dutta (1968). The inland fisheries resources of India. *J. Indian Soc. Agric. Statist.,* 20(2): 68–78.

Sunder, Shyam, H.R. Raina and K. Kumar (1979). Cause of decline and conservation of fishes in lakes. In: *Proc. Sci. Congr. 66th Session,* held at Hyderabad from 3–7 January, 1979.

Sunder, Shyam, M.J. Bhagat, C.B. Joshi and K.V. Ramakrishna (1979). On the fishery and some aspects of biology of *S. niger* Heckel from Dal Lake, Kashmir. *J. Inland Fish. Soc.,* India, 11(2).

Swingle, H.S. (1947). Commercial fish production. *Proc. Conf. South East Ass. Game Commrs.,* 8.

Chapter 19

Introduction to Various Fish Diseases

The intensification and diversification of freshwater aquaculture practices in years have led to increased incidence of fish diseases in the country. A similar picture is prevalent with regard to brackish water and mariculture system. Fisheries suffer from: a wide variety of diseases of infectious and non-infectious etiology in their natural environment. Abundant literature is available with regard to fish mortalities due to infection by viruses, bacteria, parasites, fungi, nutritional deficiencies, metabolic disorders and faulty environmental conditions like oxygen deficiency, high or very low water temperature, chemical pollutants, over crowding, malnutrition, algal blooms, etc. However, fish mortalities are not the only criterion to evaluate the effect of fish diseases. Even the morbidity which leads to weight losses and poor growth in surviving fish contribute substantial losses to the farmers. The need for maintaining healthy stock as a prerequisite to enhance fish production levels has been fully appreciated by planners and farmers alike in the recent years.

DNA-based methods have been used in diagnosis and for detection of many economically important viral pathogens of cultured finfish and penaeid shrimp. For finfish, tests have been developed for pathogens such as channel catfish virus (CCV), infectious hematopoietic necrosis virus (IHNV), infectious pancreatic necrosis virus (IPNV), spring viremia of carp virus (SVCV), viral hemorrhagic septicemia virus (VHSV), viral nervous necrosis virus (VNNV) and *Renibacterium salmoninarum* (Muroga, 1997; Plumb, 1997). DNA-based methods for detection of penaeid shrimp viruses are now used routinely in a number of laboratories around the world. DNA-based techniques will have an important role to play in efforts to develop sustainable shrimp culture in Asia and elsewhere. Production facilities in Thailand are currently using PCR techniques to screen shrimp post-larvae for white spot syndrome virus (WSSV). Culturing such larvae in closed (biosecure) or semi-closed culture systems can prevent or minimize viral infections, leading to a viable shrimp industry.

Viral Diseases

Infectious pancreatic necrosis virus (IPNV) is a highly contagious viral disease of young fish of the salmonid species. It is widely reported in Europe, Asia, America and Canada. Significant mortalities can occur from IPN outbreaks in fry. Disease signs include: Pale liver and spleen, stomach and intestine devoid of food, darker skin and trailing faecal casts. IPNV can cause an asymptomatic carrier state in salmonid fish. These carriers can shed the virus and transmit the virus to their progeny and other susceptible species. It is therefore important to detect asymptomatic carriers in order to limit the spread of infection.

Two new methods *i.e.* destructive (Gahlawat *et al.*, 2004a) and non-destructive (Munro *et al.*, 2004; Gahlawat *et al.*, 2004b) were developed for the detection of IPNV carriers in Atlantic salmon (*Salmon salar* L.) and halibut (*Hippoglossus hippoglossus* L.). These methods are easy, inexpensive and very sensitive as compared to OIE (Office International des Epizooties 2000) standard tissue culture and RT-PCR. No information was available on the longer-term persistence of IPNV in halibut and the fish were detected carriers by these methods. The destructive methods of viral isolation involve the need to kill the fish to obtain kidney and liver tissue. As halibut are repeat spawners and for broodstock testing there was a great need to develop a sensitive non-destructive method for isolation of IPNV. Macrophages are phagocytic cells and capable of killing wide range of pathogens. These play a vital role in immune response of fish and hence utilised in developing these methods.

Spring viremia of carp (SVC) is a viral disease of fish, primarily common carp *(Cyprinus carpio)* caused by ssRNA Vesiculovirus (Rhabdoviridae), known as Spring viraemia of carp Virus (SVCV) or *Rhabdovirus carpio* (RVC) (Fijan,1999). It was first described in 1971

(Fijan, 1972). Before the disease was recognised, it was variously called infectious dropsy, infectious ascites, hemorrhagic septicemia, or rubella. In Europe, the disease has had substantial impact on the production of carp, with estimated losses of 10–15 percent of 1–year–old carp or about 4,000 tons annually. In some cases, mortality rates of young carp can reach 70 per cent. This impact of the disease has led to its listing by the Office International des Epizooties (OIE) as notifiable. In Europe, it was quite some time before SVCV was adopted as a viral disease; many people thought it was caused by bacteria especially *A. hydrophila*.

The disease has been reported to occur in most European countries, the Middle East, Australia and Asia. Recently the disease has been reported from Brazil and the United States of America (Dixon, Peter EU workshop, CEFAS). The virus has also been isolated in United Kingdom from fish exported from China, although the disease has not been reported from China (Dixon, Peter EU workshop, CEFAS) and India (Asia-Pacific Quarterly Aquatic, Animal Disease Reporting System 1999–2000)

First signs of the disease may be a change in behavioral patterns of the fish. Fish may congregate in slow–flowing water, near pond banks, or lie on the bottom. Over time, the rate of respiration will decrease, as will reaction to stimulation and swimming speed. As the disease progresses, the fish become sluggish and may swim and lie on their sides.

Externally, the fish can exhibit a number of non-specific physical signs including darkening of the skin, swollen abdomen, exophthalmia (pop-eye), haemorrhages in the skin, gills and anterior eye chamber, anaemia and pale gills, and a protruding vent. Internally, the signs are dominated by building up of fluid (oedema) in all organs and in the body cavity, haemorrhages in the swim bladder, and inflammation of the intestines (Fijan, 1999).

Koi Herpes Virus

In 1988 a disease emerged that caused mass mortalities in populations of cultured Common carp (*Cyprinus carpio carpio*) and koi carp (*Cyprinus carpiokoi*). Devastating losses occurred on farms culturing carp in Israel and Germany and outbreaks of disease with similar etiology occurred in koi carp in USA, England, Germany and Netherlands. A herpesvirus was isolated from carp during many of these outbreaks that was later shown, in transmission studies, to be the cause of the disease. Outbreaks of the disease tend to be seasonal and occur at water temperature between 18 and 22°C.

The disease is highly infectious and the most prominent clinical sign is discoloration of the gills. Other signs include lethargy combined with short bouts of hyperactivity, endothalmia (sunken eyes), pale patches on the skin and an increase in mucus production on the skin and gills. Microscopic examination reveals severe hyperplasia and necrosis of the gill epithelium, and nuclear hypertrophy and margination of chromatin can be observed in some epithelial cells.

Bacterial Disease

Innumerable diseases are caused in fishes due to bacterial pathogens and several of them are reported in Indian literature. They occur in nursery, rearing and grow out ponds causing serious concern to fish farms. Some of the important bacterial pathogens are *Aeromonas hydrophila*, *A. salmonicida*, *Pseudomonas fluorescens*, *Pseudomonas* sp., *Flexibacter columnaris*, *Edwardsiella tarda*, *Vibrio alginolyticus* and *V. parahaemolyticus* which have been identified as the most commonly occurring bacterial agents of fish disease. Symptomatology, clinical pathology and anatomic pathology have been carefully correlated at the diagnosis of the diseases caused by these organisms.

Fin and tall rot in young fish are reported to be due to a mixed infection of *A. hydrophila* and *Pseudomonas fluorescens*. Bath treatment in 1:2000 copper sulphate for 1-2 minutes or painting of concentrated copper sulphate solution in the affected fish act as a good control measure. However, now a days, iodine solutions and antibiotics are used to treat fin and tail rot disease.

Dropsy is another important fish disease in India where rohu, catla and mrigal are affected mostly in the late winter and have been reported in composite fish culture ponds. The epidermis and body cavities get filled with fluid and scales protrude out from their pockets leading to severe anaemic condition. Terramycin in the feed for about 2 weeks is reported to have good effect. A mixed infection of *A. hydrophila* and *Myxosporidian* parasite is often reported in cases of *Catla catla*. Some workers in the country have attributed cases of dropsy to *A. hydrophila* in conjunction with malnutrition.

Another bacterial disease in the epidemic form has been the pop-eye-disease, the causative organism of which is identified as *A. liquifaciens* and *E. tarda*. The cornea of the eyes get vascularized leading to opacity and complete necrosis resulting into death of the fish. Corneal opacity in silver carp (*Hypophthaimichrhys molitrix* Val) and *Channa narulius* due to gram-positive bacterium, *Staphylococus aureus* has also been reported. Chloromycetin bath @S-1O mg/litre of water has been found effective in controlling the disease at an early stage by some workers.

Myxobacterial organisms have been attributed to cause a condition known as gill hyperplasia syndrome in common carp (*Cyprinus carpio*). The infectious nature of the disease has been proved by transmitting the disease in Asiatic major carp species through contact. Ulcerative disease in catla has been reported by many workers where bilateral ulcerations of the opercular bones and cranium and *A. hydrophila* isolated from the lesions. Another disease of similar nature causing external lesions over the surface of the body has been described by several workers. The causative organism has been identified as *Flexibacter columnaris*. Topical application of potassium permanganate or short bath in 500 ppm or $KMnO_4$ has been found to be very effective in completely curing the disease.

A septicaemic disease called Edwardsiellos is has been reported affecting brood fish population. *Edwsardsiella tarda* has been isolated from the disease fish showing anaemia, cutaneous lesions and gas filled abscesses in the muscle. With considerable morbidity even large scale mortality of spawn of Indian major carps has been reported in organized hatcheries. The young ones showed anatomical deformities and greyish white discoloration of the body within 48 hours post hatching and died within 72 hours. Although treatment with lodophor has been found to be effective, water quality improvement in the hatchery is the most essential component for keeping this disease away.

Epizootic Ulcerative Syndrome (EUS) has become a matter of tremendous concern not only among the fisherman and fish farmers, but also among general public, enterpreneurs, administrators and planners initially, the epizootics were mainly reported from the North-Eastern states of the country in the year 1988. Gradually, it spread throughout the country and has engulfed the entire length and breadth of the country barring only few states like the Punjab,

Jammu and Kasmir, Himachal Pradesh and Gujarat. The disease has been reported not only in the freshwater fishes, but also brackish water fishes in Orissa, Kerala, West Bengal, Tamil Nadu, Karnataka and Goa. One common feature of the disease is that it initially affects the bottom dwelling species like murrels followed by catfishes and weed fishes. Subsequently other fishes including Indian major carps are affected. Although the extent of economic loss has not been properly quantified, loss incurred by the fisheries sector due to this particular disease runs to several crores of rupees.

A number of bacteria *viz. Aeromonas hydrophila, A. punctata, Flavobacterium* sp., *Pseudomonas* sp., *E. rarda, V. parahaemolyticus* and *Streptococcus* sp. have been isolated from the affected specimens. Histopatholagical studies revealed complete loss of epidermis in the ulcerative area of the skin where the dermis and hypodermis showed characteristic granulomatous changes. Gill tissue showed hyperphlasia of the secondary lamella epithelium, fusion of lamella and clubbing of gill tips. Haemopoietic tissue of the kidney showed proliferation of macrophages. Besides bacteria, virus, fungus and parasites have also been reported to be associated with epizootic ulcerative syndrome.

Several methods have been tried or are being tried to control the disease, many antibiotics, sulfonamides, chemicals, herbal preparations, etc. have been advocated as preventive and curative measures. Lure was accepted widely among the fish farmers of the country until the formulation of a medicine named CIFAX was developed by the scientist of CIFA. Marked improvement of the ulcerative condition is noticed within 7 days of application of medicine and the ulcers are healed up within 10-14 days. Records reveal that there is no recurrence of the disease for at least eleven months in the ponds where the medicine is applied. The cost of application is only Rs. 260/- (Approx $8 US) for one ha-m water body.

Protozoan Disease

Myxosporidia are one of the most important group of pathogens capable of producing disease in fish causing heavy loss in the juveniles. Several species of this pathogen have been reported from various parts of the country causing heavy losses in intensive fish culture and fish seed rearing due to emaciation and asphyxiation.

Myxobolus cysts of varying sizes have been reported on the gills and kidney of *Catla catla*. Diagnosis of the disease can be made on the basis of gross appearance of the pin head sized greyish cysts and large number of myxosporidian spores under the microscope.

The other group of protozoa infecting fish belong to cillophora. The carp fry and fingerlings sometimes carry heavy infection of the gills and skin with *Tricodina indica* when there is excessive mucus secretion on the gills which hampers respiration and the fish start surfacing. Some species of *Tricodina* occur inside the urinary bladder. Fish affected with tricodiniasis eventually turn sluggish.. A short bath in 2–3 per cent salt solution for 5 minutes can control the infection effectively.

Whirling disease in *Cirrhirus mrigala* (Ham) has been attributed to a protozoan parasite, *Myxosoma cerebralis*. The gross changes are characterized by deformed spinal column as report in European and American salmonids suffering from whining disease. Ichthyophthiriasis or "white spot" disease has been reported causing large scale mortality and rearing ponds in different pans of the country. Presence of numerous pin head shaped white spots on the skin, fins and gills are common diagnostic features.

Helminthic Parasites

Fishes are infected on their gills and skin by many species of monogenetic and digenetic trematodes and cestodes. The important monogenetic species are *Gyrodactylus* and *Dactylogyrus*. They remain attached to the skin or gills by the help of anchors which cause wounds. Their presence cause hypersecretion of mucous and hemorrhage from blood capillaries. These wounds later get infected with pathogenic bacteria and fungi. Alternate bath in 1:2000 acetic acid solution followed by bath in 2 per cent sodium chloride solution has been found effective. Some workers have advocated dip treatment in 2 per cent salt solution followed by bath in 1:5000 formalin.

The digenetic trematodes generally infect the alimentary canal but are rarely harmful. However, at the metacercarial stage, they are harmful as they get encysted in any part of the body, *i.e.* skin, muscle, abdominal cavity, liver, eyes and brain. In these cases, fish serve as intermediate hosts while the final host is generally a fish-eating bird. In many carp fingerlings; the black spot disease is due to metacercaria of diplostomatidae. Catla and silver carp are the most susceptible species.

Cestodes are also internal parasites in alimentary canal, with higher infestations observed in catfishes than in carp. The infected fish show softening of the muscle tissue probably due to accumulation of water. In cases of heavy infestations, the abdomen bursts resulting in death of the fish. Acanthocephala have been recorded from major carps and catfishes in India, but the life history or biology of the parasites has not been studied in detail. The parasite causes lesions in the intestinal wall by its proboscis.

Parasitic copepoda are external parasites infecting skin or gills of bucco-pharyngeal cavity. Among them, species of two genera *Lernaea* and *Ergasilus* are reported from Indian carps. Lernae is reported to infect fingerlings and adults of major carps, especially catla. The parasite penetrates the muscles and reaches the blood vessels as it feeds on the host blood, rendering the fish anaemic and weak. *Ergasilu* sp. Infect the gill lamellae. They attach themselves to the gills by the help of their hook like second antenna and cause hemorrhage in the gill tissue.

Crustacean Parasites

Heavy mortality due to Argulosis or fish lice infestation have been reported in Indian major carps. Several species of *Argulus* have been reported to infect cultured fishes. These parasites inhabit the skin, fins and gills of tile host and cause extensive pathological lesions in the skin showing circular haemorrhagic patches which become ulcerated. Mucous cells proliferate and copious mucous is produced. These lesions promote secondary infections like fungi and bacteria. Rohu has been observed to be the most susceptible species among the carps.

Weekly applications of malathion @ 0.25 ppm for three consecutive doses are advised for the control. Bath in potassium permanganate@ 500 ppm for one minute once a week the combination of malathion treatment is more helpful in the eradication of the parasite.

Fungal Disease

Saprolegnia parasitica in the fry and fingerlings of major carps is characterized by a white to brown cotton like growth consisting of colonies of mycellium and filaments which appear as small to large patches on various parts of the body like fins, gills, mouth, eyes or muscle. The infection starts due to netting injury and over-crowding or lesions caused by other diseases. *Saprolegnia* often infects the fertilized eggs. They attack the dead eggs first and then spread to the viable living eggs resulting in their spoilage. Infection due to *Aphanomyces pisci* is another fungal disease where the scales become black and later fall off. The hyphae of this species are not visible like those of *Saprolegnia*.

Branchiomyces infection causes "gill rot" in fish. These pathogens invade the blood vessels of gills causing necrosis of the surrounding tissues also. Apart from these, *Aspergilus flavas*, *A. ochraceous* and Fusiformis momiliforme have been reported from some of the IRUS affected specimens.

Several different treatment measures of fungal infection in fish have been formulated, which include dip treatment in 3 per cent common salt solution, 1:2000 copper sulphate solution or 1:1000 potassium permaganate solution for 5-10 minutes. A flush treatment for 1 hour in malachite green (0.1 ppm) is found to be quite effective in curbing fungal infection.

Chapter 20

Recent Diagnostic Method for Isolation and Identification of Infectious Pancreatic Necrosis Virus (IPNV)

The intensification and diversification of freshwater aquaculture practices in years have led to increased incidence of fish diseases in the country. A similar picture is prevalent with regard to brackish water and marine culture system. Fisheries suffer from: a wide variety of diseases of infectious and non-infectious aetiology in their natural environment. Abundant literature is available with regard to fish mortalities due to infection by viruses, bacteria, parasites, fungi, nutritional deficiencies, metabolic disorders and faulty environmental conditions like oxygen deficiency, high or very low water temperature, chemical pollutants, over crowding, malnutrition, algal blooms, etc. However, fish mortalities are not the only criterion to evaluate the effect of fish diseases. Even the morbidity which leads to weight losses and poor growth in surviving fish contribute substantial losses to the farmers. The need for maintaining healthy stock as a prerequisite to enhance fish production levels has been fully appreciated by planners and farmers alike in the recent years.

Infectious Pancreatic Necrosis (IPN)

It is a highly contagious viral disease of young fish of the salmonid species. It is widely reported in Europe, Asia, America and Canada. Significant mortalities can occur from IPN outbreaks in fry. Disease signs include: Pale liver and spleen, stomach and intestine devoid of food, darker skin and trailing faecal casts. Infectious pancreatic necrosis virus (IPNV) can cause an asymptomatic carrier state in salmonid fish. These carriers can shed the virus and transmit the virus to their progeny and other susceptible species. It is therefore important to detect asymptomatic carriers in order to limit the spread of infection.

Two new methods *i.e.* destructive and non-destructive were developed for the detection of IPNV carriers in Atlantic salmon (*Salmon salar* L.) and halibut (*Hippoglossus hippoglossus* L.). These methods are easy, inexpensive and very sensitive as compared to OIE (Office International des Epizooties 2000) standard tissue culture and RT-PCR. No information was available on the longer-term persistence of IPNV in halibut and the fish were detected carriers by these methods. The destructive methods of viral isolation involve the need to kill the fish to obtain kidney and liver tissue. As halibut are repeat spawners and for broodstock testing there was a great need to develop a sensitive non-destructive method for isolation of IPNV. Macrophages are phagocytic cells and capable of killing wide range of pathogens. These play a vital role in immune response of fish and hence utilised in developing these methods.

Destructive Method

The macrophages were isolated following Secombes (1990) with some modifications. Head kidney (HK) from Atlantic salmon and halibut was dissected using sterile conditions into ice chilled 3ml cell separation medium. This medium consists of Leibovitz (L-15, PAA Laboratories) medium, supplemented with heparin 10U/ml, 2 per cent FCS, 50U/ml penicillin and 50 µg/ml streptomycin (P/S). The HK was passed through a 100 µm nylon mesh and left on ice for 10 min. The cell suspension was carefully layered onto a 34/51 per cent discontinuous Percoll density gradient. The gradients were centrifuged at 540g for 35 min at 4°C with low brake. The band of cells at the gradient interface was collected and diluted 3-4 times with low serum medium (L-15 medium containing 0.1 per cent FCS, P/S) and centrifuged at 540g for 15 min at 4°C. The supernatant was discarded and the pellets were resuspended in the low serum medium to eliminate the Percoll gradient and then centrifuged again as before. The supernatant was discarded and the cells were resuspended in 1 ml L-15 containing low serum medium. The cells were counted using a haemocytometer, their viability assessed by trypan blue exclusion and adjusted to approximately 1 × 10^7 cells/ml in low serum medium. Aliquots of 100 µl were added to each well of a 96-well tissue culture plate (Corning Incorporated) and cells were left to attach overnight at 22°C. The Percoll is very costly and this method is time consuming and not feasible for large number of samples. So we have discarded the percoll gradient step and isolated the adherent macrophages from head kidney. The HK material (1 gram) passed through a 100 µm nylon mesh using 2 ml of cell separation medium and 100 µl of the cell suspension was added directly into each well of a 96-well tissue culture plate and cells were left to attach overnight at 22°C. To check whether this method is applicable for field samples, head kidney material was divided into 4 sections, transferred into cell separation medium and stored in an ice-packed polystyrene box. Macrophages were prepared after 0, 24, 48 and 72 hours of sampling. The non-percoll macrophages were compared with percoll macrophage for isolation of IPNV and were found similar results. This method was also compared with OIE standard tissue culture and RT-PCR and found more sensitive as compared to both.

Non-destructive Method

Blood was collected aseptically from the caudal vein of anaesthetised fish in heparinised vacuatte tubes (Greiner bio-one). The blood (0.5ml) was diluted three times with cell separation medium. The cell suspensions were layered onto 34/51 per cent Percoll discontinuous gradients. The gradients were then centrifuged at 540g for 35 min at 4°C with low brake. The band of cells at the gradient interface were collected and diluted 3-4 times with low serum medium and centrifuged using the same conditions as before. The supernatant was discarded and the pellets were resuspended in 1ml low serum medium. The cells were counted using a haemocytometer, their viability assessed by trypan blue exclusion and adjusted to approximately 1 × 10^7 cells/ml in low serum medium. Aliquots of 100 µl were added to each well of a 96-well tissue culture plate and cells were left to attach overnight at 22°C.

Whole blood or blood diluted three times with cell separation medium was aliquoted (100 µl/well) into a 96-well tissue culture plate and cells were left to attach overnight at 22°C. In order to check whether this method is feasible for field samples, the blood was kept in heparinised vacuatte tubes and stored in an ice-packed polystyrene box. Blood leucocytes were prepared after 0, 24, 48 and 72 hours of sampling and compared with earlier methods. Sometime the fish is very small or not able to take out enough quantity of blood, different quantities (10, 50, 100, 250, 500 µl) of blood was diluted with cell separation medium and aliquoted (100 µl/well) into 96 well plate, incubated overnight and compared with earlier methods.

Inoculation of Samples onto CHSE-214

After overnight incubation of kidney and blood leucocytes, the non-adherent cells were gently resuspended using a pipette and the supernatant containing the non-adherent cells from two wells was inoculated into CHSE-214 25cm^2 flasks. The adherent macrophages and blood leucocytes were washed 2-3 times with L-15 medium supplemented with 5 per cent FCS and P/S. After one hour two wells of each macrophages and blood leucocytes were lysed for 15 min with 10 µl Trypsin-EDTA lysis buffer (Invitrogen) per well. Any remaining adherent material was scraped off the bottom of the well and the well contents were passed into CHSE-214 25cm^2 flasks. Cultures were incubated at 15°C for seven days. The sample and control flasks were examined for the presence of CPE on days 3–7 after incubation and positive confirmation of IPNV infection was obtained using an IPNV-ELISA. The strain IPNV SP Spjarup (EU Reference Laboratory) was used for virus positive controls during each set of tests. Negative control CHSE-214 flasks were separately inoculated with 100 µl of each of the reagents used in the processing of macrophages (cell separation medium, low serum medium, high serum medium, 34/51 per cent Percoll or trypsin/EDTA lysis buffer).

Macrophage Culture

Preparation of Solutions

(i) L-15 medium with 2 per cent FCS

500 ml L-15 without glutamine

10 ml FCS (2 per cent)

5 ml P/S (Penicillin and Streptomycin)

0.5ml Heparin (10 unit/ml)

(ii) L-15 medium with 0.1 per cent FCS

500 ml L-15 without glutamine

0.5 ml FCS (.1 per cent)

5 ml P/S (Penicillin and Streptomycin)

(iii) L-15 medium with 5 per cent FCS

500 ml L-15 without glutamine

25 ml FCS (5 per cent)

5 ml P/S(Penicillin and Streptomycin)

(iv) Percoll solutions

51 per cent Percoll

51 ml Percoll

10 ml x10HBSS

35 ml DW(deionised) first

Add–N HCl to adjust pH 7.2-7.4

Add more water to make 100ml

34 per cent Percoll

34 ml Percoll

10 ml x10HBSS

50 ml DW(deionised) first

Add–N HCl to adjust pH 7.2-7.4

Add more water to make 100ml

Procedure

Using sterile technique, dissect out the head kidney (HK) from a 500-700g fish

Take out the HK and cut into 3-4 pieces and pass through 100 µl nylon mesh with L-15 medium containing 2 per cent FCS, 100 µl/ml P/S and 10 µl/ml heparin

Take 5 ml of 51 per cent percoll in 6 vials of 15ml and add 3 ml of 34 per cent percoll slowly not to disturb the other layer (Can be prepared in advance)

Add 1 ml each of cell suspension in all the vials

Centrifuge at 1600rpm for 25-30 min at 4°C

Discard the upper layer of RBC's

Take a band of cells in between percoll layers

Add 3-4 times L-15 supplemented with 0.1 per cent FCS and 100 µl/ml P/S and centrifuge at 1600rpm for 15-20 min at 4 °C to get rid of percoll

Discard the supernatant and add 3-4 times L-15 and centrifuge at 1600rpm for 15-20 min at 4°C to get rid of percoll

Discard the supernatant and add 1ml of L-15 to the pellets and resuspend it into all the 6 vials (Only 1ml of L-15 is to be added in all the vials)

Take 2-3 µl and dilute it 25 times and add 2 µl of trypan blue

Count the cells in Haemocytometer under microscope

Calculate the dilution factor by using the formula given below and add the required vol. of L-15 and adjust the cell suspension to 1×10^7 or 2×10^7 viable cells/ml

Distribute it into 100 µl of this suspension in 96 well plate

Formula

$$\frac{\text{No. of cells} \times \text{No. of squares} \times 10^4 \times \text{Dilution factor}}{1 \times 10^7 \text{ or } 2 \times 10^7}$$

$$\frac{40\ (e.g.) \times 5 \times 10000 \times 25\ (e.g.) = 50000000}{10^7 = 10000000} = 5$$

i.e. So you have to add 5 ml in one ml cell suspension and distribute it into 60 wells (100 µl each)

Infect the one day old macrophages (96 well plate) with IPNV Cole deep virus (851/99, Batch 2) from 10^{10} to 10^1.(If the medium is 100 µl then add 10 µl; if 150 µl then add 15 µl; and start from 10^{10} to 10^1). Keep three cells as control and make a replication of three wells for infection

Incubate the cells at room temperature

Take out the supernatant in Cryovials next day

Add 500 µl in CHSE flasks (cells should be 50 per cent grown) containing 5 ml media (1:10)

Incubate the cells at 15°C for 7 days

Now add 10-20 µl of Trypsin EDTA in each well for 2-3 min. The cells starts dissolving in EDTA

Add 30-40 µl of L-15 medium

Take out the supernatant in bijou tubes and sonicate the sample @75 per cent for 1 min

Now add this supernatant in CHSE flasks containing 5 µl cell suspension

Incubate the samples at 15°C for 7 days

Enzyme-linked Immunosorbent Assay (ELISA) for IPNV

The method used was a double antibody sandwich ELISA using Rabbit anti-IPNV polyclonal antiserum to capture IPNV in the tissue culture supernatants and CE4 monoclonal antibodies (MAb) specific to IPNV followed by a goat anti-mouse alkaline phosphatase conjugate were used for IPNV detection. The protocol has been described in detail by Smail *et al.* (2003).

Titration of Virus from Macrophages

Following overnight incubation of kidney cell suspensions, with no prior fractionation on Percoll, in 96 well plates, the supernatant containing non-adherent cells was discarded and the adherent macrophages were washed and cultured in L-15, 5 per cent FCS, P/S. After 1 h in culture, the supernatants were discarded and the macrophages were lysed in Trypsin/EDTA buffer and serial dilutions of the macrophage lysates were inoculated into a 96-well CHSE–214 cell culture plate and observed for the development of CPE. The $TCID_{50}$/100 µl was calculated seven days after inoculation.

Reverse Transcription Polymerase Chain Reaction (RT-PCR) for VP2 Gene for Isolation of IPNV

The reverse transcription-polymerase chain reaction for detection of IPNV was performed following Taksdal *et al.* (2001) with minor modifications on Atlantic salmon and halibut samples. RNA was extracted from whole kidney samples using TRIzol® reagent (Invitrogen) according to the manufacturer's recommendations. RNA was resuspended in DEPC-treated ddH_2O. Concentration and purity was quantified by measuring absorbance at 260 and 280nm (GeneQuant II, Pharmacia Biotech). In the RT step 2 µg RNA was denatured for 10min at 55-60°C followed by cooling on ice. The reverse transcriptase mix (10.5 µl) was added, giving a final volume of 20 µl containing 1 × 1st strand buffer (Invitrogen), 10mM of each deoxynucleoside triphosphate (Invitrogen), 2.5mM of random primer (Promega, Inc), 0.1M DTT (Invitrogen), 20U RNaseOUT™ (Invitrogen), and 200U Moloney Murine Leukemia virus reverse transcriptase (MMLV-RT) (Invitrogen). This mixture was incubated for 1 hour at 37°C.

Five µl cDNA were added to 45 µl of PCR mix, giving final concentrations of 1x Reaction Buffer (Bioline, London, UK), 1.5mM $MgCl_2$, 0.2mM deoxynucleoside triphosphates (each), 28pmol of each primer, 2U of *Taq* polymerase (Bioline) and DEPC-treated ddH_2O. The reactions were overlaid with mineral oil (Sigma). The primers used for IPNV detection were IPNV 1(forward, 5'-ATCTGCGGTGTAGACATCAAAG-3') and IPNV 2(reverse, 5'-TGCAGTTCCTCGTCCATCCC-3'). Amplification reactions were subjected to 35 cycles of 94°C for 1min, 60°C for 1min and 72°C for 1min followed by an extension step of 5min at 72°C. The PCR products were subjected to electrophoresis on a 1.5 per cent (w/v) gel in TAE buffer (40mM Tris, 20mM acetic acid, 2mM EDTA), stained with ethidium bromide and photographed using the Syngene GeneGenius Gel Documentation and Analysis system. (GeneSnap software version 4.0).

Chapter 21

Introduction to Freshwater Prawn (Scampi), *Macrobrachium rosenbergii* Diseases

It is believed that freshwater prawn (scampi), *Macrobrachium rosenbergii* is less susceptible to diseases than penaeid shrimp (*Panaeus monodon*). It might be as a result of lower stocking densities used in culture practices and less transfer of brood stock. However, a variety of diseases have been reported in larval, juvenile and adult scampi, which include fouling protozoans such as *Zoothamnium, Vorticella* and *Epistylis*; fungal pathogens such as *Lagenidium, Sirolpidium* and *Fusarium*; bacteria such as *Vibrio, Aeromonas, Leucothrix* etc., and also non-pathogenic diseases such as stress and neurotic symptoms, moult failure and so on. This may be due to shorter life cycle and faster metabolism of these animals, which results in weaker resistance to pathogens. Prawns are highly sensitive to pathogens during molting period when they cannot feed, have a week body and poor mobility. The incidence of disease is higher in low quality water, particularly if oxygen levels are low. Prawns need good quality water and enough plant life. The following diseases of freshwater prawn were reported by various workers:

Larval Midcycle Disease

It is widely occurs in Thailand, Philippines, Malaysia and India. The larval stages IV to XI are usually found to be affected from this disease. The symptoms appeared about 10 days post-hatching and high larval mortality was observed. The weak larvae stop feeding and settle at the bottom of the rearing tank and exhibit spiral swimming behaviour along with vertical axis turning. The larvae become bluish grey in colour and the hepatopancreatic epithelium get atrophied

Possible Causes

This disease might be due to nutritional deficiency, since lecithin deficiency is often said to occur along with this disease. Bacteria, *Enterobacter aerogens* has been reported to be associated with this disease.

Preventive Measures

Infected stock should be immediately destroyed and tanks should be disinfected and dried. The farmers should adopt standard hatchery techniques that may prevent vertical or horizontal transmission of this disease.

Bacterial Diseases

(*a*) Bacterial Necrosis

This is a common disease observed in larvae, post-larvae or adults in freshwater prawns. It is variously termed as 'black spot', 'brown spot', 'burnt spot', 'Shell Disease' or Chitinolytic bacterial disease. This is caused by the invasion of chitinolytic bacteria, which break down the chitin of the exoskeleton, leading to erosion and melanization (dark brown to black pigmentation) at the site of infection. Several chitinolytic bacteria (Gram- ve rods) such as *Vibrio* spp., *Pseudomonas* spp., *Aeromonas* spp., *Spirillium* spp. etc. are involved. This disease reduces the value of the harvested prawns, apart from causing mortalities.

This disease can be controlled in captive and cultured populations by reducing over crowding, proper husbandry and system hygiene.

(*b*) Filamentous Bacterial Disease

Filamentous bacteria such as *Leucothrix muco, Thriothrix* spp., *Flexibacter* spp. etc. sometimes cause mortalities in freshwater prawns by discolouration of gills and associated secondary infections. The larvae or the post larvae become moribund, with reduced mortality, poor feeding and growth.

Increasing husbandry and hygiene standards will improve the situation.

(c) Luminescent Bacterial Disease

Luminescent bacteria especially *Vibrio harvey* often infect the prawn larvae reared in hatchery conditions, when they appear in seawater. They spread very fast and poor hatchery conditions increase their virulence. The infected larvae become fluorescent in dark light, with reduced feeding and in severe cases mass mortalities occur.

Treatment

This disease can be cured by using specific antibiotics and disinfecting the pond water.

Fungus Disease

Larval mycosis due to *Fusarium* spp., *Lagendium* spp., *Sirolpidium* spp. cause severe mortalities to prawn larvae often as a secondary infection. *F. solani, F. moniliformes, L. callinectes* etc. are commonly found. Fungus infects the dead or damaged tissue caused by wounds or other infections resulting in locomotory difficulties due to mycelial growth. In serious infections of *Lagendium* extensive non-septate highly branched mycelia invade through out the body replacing all the tissues. Specialized hyphae protrude through the cuticle.

Viral Disease

Scampi is believed to be free of viral diseases a case of occurrence of white spot virus has been reported from Taiwan in *M. rosenbergii.* Signs closely resembling white spot syndrome have been observed on its exoskeleton as reported by Peng *et al.* (1998).

Protozoan Parasites

Fouling protozoans such as *Zoothamnium, Vorticella* etc. are of common occurrence in larval, post larval or adult scampi. They attach to the body and appendages and disrupt mobility and feeding. Mortality occurs only in severe cases of susceptibility. Formalin treatment will be helpful to save the affected.

Other Diseases

(*a*) White Muscle Disease

Opacity or 'white muscle' is a condition to the prawn arising out of any stress such as over crowding, extreme temperature or pH, low dissolved oxygen etc. The body appears opaque white proceeding forward from the tail region. The condition is often reversible once the stressed condition is corrected.

(*b*) Blackening of Gills

Blackening of gills of scampi may sometimes be found as a manifestation of several other disease syndromes, precipitation of dissolved chemicals, turbidity, vitamin C deficiency etc. General discolouration of gills may occur due to melanization of tissue and necrosis, which may be visible through the side of carapace.

Chapter 22

Ectoparasitic Diseases in Freshwater Fishes and their Control

An overview of the global fisheries trends brings out the growing important of aquafarming in recent years. Indian freshwater aquaculture has evolved from the stage of domestic activity in West Bengal, and Orissa to that of an industry in recent years with states like Andhra Pradesh, Punjab, Haryana and Maharashtra taking up fish culture as trade. Fish has long been an important source of food for people all over the world. The importance of fish as a source of high quality, balanced and easily digestible proteins, essential amino acids and nutraceuticals like PUFA is well understood. According to an FAO survey, the number of people involved in fishing and fish farming worldwide has more than doubled since 1970. Most of this growth has taken place in the 1980's and mainly in Asian countries where four fifths of the world fish farmers dwell. In India, the importance of fisheries sector is well known as it provides of livelihood to seven million farmers, prime mover of coastal economy and a major foreign exchange earner. Development of several suitable technologies coupled with extension activities during the last five decades has pushed Indian fish production many fold, particularly to become second largest aquaculture fish producer and a thousand fold increase in the value of marine products exports. Presently the contribution of fisheries to the GDP of the country is estimated at 1.4 per cent whereas the export earnings of fisheries products reached Rs. 6,443 crores.

The country is endowed with vast and varied resources possessing river ecological heritage and rich biodiversity. The share of inland fisheries sector, which was 29 per cent in 1950-51 has gone up over 49 per cent in 2001 indicating increasing contribution of the inland sector to the total fish production.

Table 22.1: Inland Fishery Resources

Sl.No.	*Resources*	*Area (hecs)*
1.	Ponds and tanks	2.36 million ha
2.	Oxbow lakes and derolect water bodies	1.3 million ha
3.	Reservoires	2.05 million ha
4.	Rivers	45,000 km
5.	Canals	1,26,334 km
6.	Esturies	2.7 million ha
7.	Brackish water	1.42 million ha

Aquaculture has become the fastest growing food production system with a growth rate of over 9 per cent in the last 10 years. Considering the increasing demand of fish to meet the nutritional requirements of the growing population in the country and the potentials of the fisheries sector in general and freshwater aquaculture in particular with regard to the contribution to the country's food basket, several technologies have been developed and evaluated for different agro-ecological situations for increasing fish production in the country.

Dictionary defines disease as the deviation from the state of normal health. Every living organism are prone to several diseases, and fish makes no exception. In aquatic ecosystem there are so many pathogens are present that causes various diseases of culturable fishes *viz.* virus, bacteria, fungi and other parasites. Parasites are those micro or macro organism that live on the surface of the host or embeded inside the body cavity. These are called ectoparasite and endoparasite, respectively. Ectoparasites are very much important as concern to the fish disease. The wound or injury caused by the ectoparasite are the open invitation of other pathogens like fungi, virus, bacteria etc.

Fish under normal condition is in a state of equilibrium with its environment and disease producing organism. Any alteration in the environment disturbs this equilibrium resulting in stress to the fish and they become more vulnerable to disease. Wedemeyer and McLeay (1981) is defining the relationship of environment vis-à-vis pathogen vis-à-vis fish explained that the presence of the fish pathogens will result in the epizootics only if unfavourable environmental condition also exist and the host defence mechanism has been compromised. Intensification of aquaculture, especially in the absence of adequate knowledge about environmental hygiene balanced nutrition and host-pathogen relationship has resulted in the occurrence and spread of economically serious diseases in the industry. Disease outbreaks cause significant losses in aquaculture production and are affecting economic development.

This chapter include the common ectoparasitic diseases of culturable freshwater fishes in India and their management practices. Followings are the ectoparasites that causes various diseases in freshwater fishes.

Table 22.2

	Groups of Parasite		*Name of the Parasite*
1.	Protozoa	i.	*Ichthyopthirias* spp.
		ii.	*Trichodina* spp.
		iii.	*Myxobulus* spp.
		iv.	*Costia* spp.
2.	Crustacea	i.	*Argulus* spp.
		ii.	*Ergasilus* spp.
		iii.	*Lernea* spp.
3.	Helminth	i.	*Dactylogyrus* spp.
		ii	*Gyrodactylus* spp.
		iii	*Diplostomum* spp.

Now different ectoparasitic diseases caused by the above parasite and their common symptoms discussed in Tables 22.3–22.4.

Table 22.3

Name of the Disease	Causative agent	Most Affected Species	External Symptoms
1. White spot or ich diseases	*Ichthyophtheries multifilis*	Fry and fingerling of *Catla catla, Labeo rohita, Cirrhinus mrigala* and other carps.	1. Minute white spots are visible on the surface of fish. 2. Rubbing against hard object may be common. 3. Attachment tissue become redish.
2. Trichodiniasis	*Trichodina reticulata, T. niger*	Fry and fingerling of *Catla catla, Labeo rohita, Cirrhinus mrigala* and other carps in nursery and rearing ponds.	1. Colours of the gill turn pale. 2. Fish secrete excessive amount mucus. 3. Respiratory function with hamper.
3. White scale spot disease	*Myxobulus sp.*	All most all culturable fishes can attacked by this parasite	1. White cysts are found on surface of scale and body of fishes. 2. Scale become loose and perforated in many places. 3. Ulcer may develop at the attachment site.
4. White gill spot disease	*Myxobulus sp. Thelohanellus sp.*	All the carps (major, minor Chinese carp) and other freshwater fishes.	1. White cysts are visible on the surface of the gill. 2. Excessive mucus secretion occur. 3. Gill lamellae will be damaged. 4. Surfacing of fishes may be common.

Contd...

Table 22.3–Contd...

Name of the Disease	*Causative agent*	*Most Affected Species*	*External Symptoms*
5. Costiasis	*Costia necatrix*	All the freshwater fishes are prone to attack by this parasite	1. Colour of the body of fish fade. 2. Fish become restless and rubbing against object. 3. Fin may damaged. 4. Surfacing of fishes are common.
6. Argulosis (fish louse)	*Argulus* sp.	*Catla catla, Labeo rohita, Cirrhinus mrigala* and other carps.	1. Adult parasite visible moving freely on the surface of the fish. 2. Fishes become restless with erratic swimming. 3. Scap against hard parts. 4. Attachment site become ulcerated.
7. Lernarosis (anchor work)	*Lernaea sp.*	All the major carps, common carp and Chinese carps are attacked by the parasite	1.Fishes become restless and scrap against object. 2. Fishes show erratic swimming behaviour. 3. Skin redden many places. 4. Attachment site show ulceration.

Contd...

Table 22.3–Contd...

Name of the Disease	*Causative agent*	*Most Affected Species*	*External Symptoms*
8. Ergasilosis	*Ergasilus sp.*	All the freshwater fishes can be attacked by the parasite	1. Infection occur in gill, buccal cavity and fin. 2. Parasites are look like white bodies less than 2 mm long. 3. Gill tissue damaged. 4. Surfacing of fishes.
9. Dactylogyrosis (Gill fluke)	*Dactylogyrus sp.*	Fry and fingerling of carps	1. Fading of normal colour of gill. 2. Fishes are restless and scrap against object. 3. Gill moves rapidly. 4. Surfacing may common
10. Gyrodactylosis (skin fluke)	*Gyrodactylus sp.*	Fry and fingerling of carps	1. Colour of the body fade. 2. Dropping of scale. 3. Scrap against object due to irritation. 4. Excessive mucus secretion occur.
11. Black spot disease	*Diplostomum sp.*	All small and adult carps can attack by the parasite	1. Black ovid patches on the surface. 2. Cysts vary from few to hundred.
12. Leech infection	*Pecicola sp.*	All the freshwater fishes are attacked by the parasite	1. Infection occur on skin, gill and fins. 2. Hemmorhage found on the site of attachment. 3. Fishes become restless and rubbed against hard part.

Table 22.4: Treatment

Name of the Disease	Treatment Recommended	Dose Applied
White spot disease	1. Formalin bath treatment	200 mg/L for 1 hr.
	2. Malachite green in pond	1.5 mg/L
	3. Methylene blue	1-3 mg/L
	4. $KMNO_4$ bath treatment	4 mg/L for 1 hr.
	5. NaCl bath treatment	3 per cent for 1 hr.
Trichodiniases	1. Maintain the water quality by adding freshwater	—
	2. Decrease the stocking density	—
	3. NaCl bath treatment	2-3 per cent for 1 hr.
	4. $KMNO_4$ in pond	4 mg/L
	5. Formalin in pond	25 mg/L
White scale spot disease	1. Decrease the density of fish	—
	2. Mohua oil cake in pond	25 mg/L
	3. NaCl bath	3-5 per cent for 30 min for 1 hr.
Costiasis	1. Acetic acid	2-5 per cent
	2. $CuSO_4$ bath treatment	1-4 mg/L for 3-5 min.
	3. Formalin dip treatment	400 mg/L
	4. Malachite green dip treatment	60 mg/L
	5. $KMNO_4$ bath treatment	4 mg/L for 1 hr.
	6. NaCl dip treatment	3-5 per cent
Crustacean disease (Argulosis, Lernaeosis, Argasilosis)	1. NH_4Cl bath treatment	1000 mg/L for 4 hrs.
	2. Benzine hena chloride (BAC)	0.12 mg/L
	3. Masoten	0.25 mg/L
	4. NaCl bath treatment	3-5 per cent for 1 min.
	5. NaCl in pond	0.8-1.1 per cent for 3 days
	6. Diflubenzuron	0.01 mg/L

Contd...

Table 22.4: Control

Name of the Disease	*Treatment Recommended*	*Dose Applied*
Dactylogyrosis and Gyrodactylosis	1. Acetic acid dip	5 per cent for 1 min
	2. Formalin bath treatment	250 mg/L for 1 hr.
	3. Malachitegreen bath	5 mg/L for 1 hr.
	4. Mebendazole bath	0.01 mg/L for 24 hrs.
	5. Methylene blue	2.9 mg/L
	6. Masoton	0.25 mg/L
	7. $KMNO_4$	2.4 mg/L
	8. NaCl bath treatment	3-5 per cent for 1 hr.
	9. Tricholorfon bath	1.6-0.4 mg/L for 24 hrs.
Black spot disease	1. Pond should be free from molluscan population	—
	2. Prevent entering of bird	—
	3. Pond may be dried	—
	4. Picric acid bath	2-7 : 100000 for 1 hr.
Leech infection	1. NaCl bath treatment	3-5 per cent for 10-15 min
	2. Masotin in pond	0.5-1.0 mg/L
	3. Glacial acetic acid dip	1 : 1000 for 1-2 min
	4. Adult parasite can be pull out with the help of a forcep	
	5. Pond may be dried and treated with chlorinated line	

Chapter 23

Hazards of Fish Culture

Like other farming enterprises fish farming runs the risk of encountering several incidental hazards, which may cause heavy losses unless they are anticipated and remedial measures taken in time to overcome there. Most of the problems emanate because of poor management, excessive growth of weeds, appearance of algal blooms, pollution of water, in grass of unwanted species of fishes, shrimps and insects and depletion of oxygen all create serious problems resulting in loss of fish.

Fishery Management

Fishery management is the application of scientific knowledge of fish life to the problems of fish production. The primary objective of management of fish farming is to increase the fish production in pond. Production is three types:

1. Quantitative production–Largest possible quantity of fish without any regard to the quality of fish.
2. Qualitative production–Means producing high quality fish.
3. Economic production of aims at producing the maximum possible quantity of fish having high market value.

The production of fish in a pond can be increased by correct choice of fish, improvement of pond, artificial or supplementary feeding and rate of stocking.

Excessive Growth of Weeds

Control of weeds in a fish culture pond is aimed at reducing the excessive growth of weeds to a level at which they become beneficial and not merely harmless. It must be kept in mind that a moderate growth of weeds is not only useful but essential for the pond productivity. Three kinds of methods are used for the control of weeds in fish culture ponds *viz.,* physical, chemical and biological.

Physical Methods

Draining of Pond

A drainable fish pond is empted and exposed to the strong summer sun for several days.

Manual Cleaning

Cutting of emergent and floating-leaved weeds and hand picking of cut parts and floating weeds is possible in shallow ponds only.

Mechanical Cleaning

Deep ponds are cleared of weeds with mechanical devices such as bottom lakers, weed-cutters, winches and barbed wire.

Chemical Methods

Many chemicals are known to destroy the weeds. These are called herbicides.

1. 2,4-D is effective against water hyacinth, lotus, sedges and creepers. Its use prevents regrowth of weeds for a long time, if not permanently.
2. Diuron is effective against submerged weeds.
3. Ammonia is effective against all aquatic weeds and the perilinium bloom is varying doses. Besides destroying weeds, ammonia also adds nitrogen to the pond.
4. Copper sulphate (3 ppm) over several days control algal blooms.
5. Addition of lime or phosphate may change pH of water to a level harmful for blue green algae.
6. Chrotozyphos at a concentration of 0.1 to 1 mg/litre is effective in the control of zooplankton blooms.

Biological Controls

Algal Blooms

Lamina controls algal blooms by shading *i.e.* preventing sunlight. Phytoplankton phagous fish, such as silver carp can keep algal blooms under check. Seasonal blooms of phytoplankton are responsible for various colours of the water such as green, dirty brown, yellowish green, yellowish brown, reddish brown etc. The species forming blooms are pumdonia, volvex, euglena, blue-green algae, microcystis, anabaena of oscillatoria etc. Algal blooms usually make their appearance after treatment of the land with heavy loses of inorganic fertilizers like urea and superphosphates. Bloom also appears after treatment with weedicides like 2,4-D, sodium arsenite etc. Algal blooms are responsible for giving a disagreeable taste and smell to the water. They cause heavy mortality of the fish due to choking of the gills, depletion of oxygen and liberation of toxic substances.

Aquatic Weeds

Chinese grass carp can clear the fish pond of all aquatic weeds, except water hyacinth. Herbivorous birds such as swan, if introduced into a fish pond, will consume submerged vegetation to a considerable extent. Addition of fertilizers, mostly organic such as compost and cowdung, causes a rich growth of phytoplankton which cuts off sunlight to deeper waters and hereby checks weed growth. When we use the combined methods they are more effective in weed control.

Pollution of Water

The term pollution broadly refers to any change in the natural quality of environment brought about the chemical, physical or biological factors. Environment pollution is unfavourable change of our surroundings due to direct or indirect activities of man. Domestic wastes, sewage and industrial effluents are generally allowed to add to the rivers without any pretreatment, causing pollution of water. The waste products of some of the industries are extremely poisonous to fish life and cause depletion of fish population by adverse changes in the physical, chemical and biological properties of freshwater. In addition to the industrial and domestic wastes, a large number of agricultural pesticides and insecticides have further increased the hazards of pollution of water.

Pollution through Industries

In India, pollution of river water takes place at various centers of industrialization, chiefly at Bombay, Calcutta, Madras, Kanpur, Delhi and Sindri, where there is a heavy concentration of factories and mills. At Bombay, a large number of industries dealing with aluminium, steel and iron, textiles, chemicals, synthetic goods, petro-chemicals, pesticides etc. discharge their highly acidic wastes into the canals and river, thus greatly affecting the fishes. At Calcutta, large number of industrial complexes is located on both sides of the river Hoogly and discharges their wastes into it. The main pollution sources are the wastes from paper and pulp, textiles, distillery and chemical factories. The effect of pollutants extends to several kilometers, from the site of discharge, reducing the plankton and other food of fishes in the affected area. In addition to industrial wastes, domestic sewage from the enormous Calcutta population is discharged into the river Kulti, causing unfavourable affects of fish fauna in 20-30 kms area. This reduces the plankton and other organisms considerably, so that several fish species are virtually eliminated.

DDT and chloral hydrate are the chief pollutants of the river Jamuna at Delhi. A number of sugar factories, distilleries, soap and paint industries and paper factories located at various centers in Uttar Pradesh, discharge their wastes into the river causing heavy pollution of water. The discharge contains alkalis, ammonia, cyanide, chromates and phenols. Finally suspended coal pesticides and fly ash further add to the pollution, so that large scale fish mortalities have been reported.

Pollution through Domestic Sewage

In India domestic sewage from most of the big cities is allowed to be discharged into the rivers without any treatment to render it harmless. Only few cities possess facilities for even partial or primary treatment. The river water becomes considerably polluted in the nearby areas. Untreated sewage and laundry detergents coming from the household when added to the rivers, increase the amount of sulphates, nitrates and chlorides considerably. More compounds favour the growth of algae producing water bloom which consume most of the oxygen from water and cause toxic effects.

Pollution by Fertilizers and Insecticides

The use of pesticides to control agricultural parts is now increasing every day. Several kinds of insecticides like DDT are also used large quantity to control diseases like malaria and filarial. These chemicals are poisonous and are washed into the rivers, tanks and pond endangering fish life. The important pesticides and insecticides in common use are DDT, aldrine, benzene hexachloride, endrex etc. chemical fertilizers used in the agricultural fuels like urea and superphosphates are washed down into the fishery ponds and nursery tanks and prove to be harmful for all sorts of organisms.

Effects on the Fishes

Different types of pollutants effects directly or indirectly as mentioned below:

1. There is increase in the osmotic pressure.
2. The water becomes acidic and pH lowered below 4.
3. They bring about oxygen depletion.
4. The toxic substances injure the mucous membrane of the gills and prone to be fatal, by affecting respiration.
5. The suspended matter injures the gills causing mechanical damage to the soft parts.
6. Pollutants reduce the zooplankton and phytoplankton and other aquatic organisms, thus indirectly affecting the fish by food depletion.
7. The spawning grounds and nesting material are destroyed, preventing natural spawning of fishes.
8. The fish flesh gets foul odour due to certain pollutants, and become unpalatable.
9. Sewage and silt bring about an early ageing of the labs and ponds.

Different chemicals, pesticides, sewage waste are affecting different part of the fish and increased the mortality. It is difficult to estimates the exact effect of pollution and the problem needs careful laboratory and field experiments, but there is no doubt that various changes brought about in the ecological conditions of the environment prove harmful and even fatal to the fishery of the region.

Unwanted Species

It is always desirable to guard against undesirable fishes, insects and diseases which adversely affect fish growth and production. However, over-crowding, shortage of food, unhealthy conditions and excessive handling of the fish should be avoided as far as possible. Oyster and snakes also do considerable damage to the fish crop by feeding on fry and fingerlings. Molluscs when present in large numbers adversely affect fish culture. They can be controlled by stocking the fish *Pangasius pangasius* in the pond. They feed on mollusks and reduce their infestation.

Depletion of Oxygen

Dissolved oxygen is most important for the animal and plant life in a pond. Oxygen requirement for aquatic varies in different cultivable fishes. Carps need 6 to 7 mg per litre concentration of dissolved oxygen in water at 20 °C. The factors which reduce the level of dissolved oxygen in water include cloudy days, higher temperature, higher salinity, decomposing organic matter, sewage, fertilizer and manure. There is diurnal variation also. Oxygen concentration generally increased during the day due to photosynthesis by pond flora and decreases at night due to respiration by pond flora and fauna. Hence, excessive growth of algae and submerged weeds in a fish pond should be checked. Excess of oxygen due to over carbon assimilation causes mortality due to over metabolism. Deficiency of oxygen, especially during late hours of the night, is fatal to the fish. On cloudy days, photosynthesis is reduced and causes oxygen deficiency at night. Oxygen deficiency in labs and ponds causes migration, attack of parasites, fungus diseases and death due to suffocation.

Index

The Authors

Dr. S.K. Gahlawat, born in 1968, is an Assistant Professor in the Department of Zoology and Aquaculture, CCS Haryana Agricultural University, Hisar. He did his M.Sc. (1989), M. Phil (1990), Ph.D. (1992) in Zoology from Kurukshetra University and Post-doctoral research in fish biotechnology/immunology from FRS Marine Laboratory, Aberdeen, UK (2004). He has about 40 publications to his credit in various journals of national and international repute. He has participated in a number of refresher courses, training programmes, workshops, seminars, symposia and conferences at national and international levels on different aspects of science, thus gathering vast knowledge of current developments in aquaculture.

Dr. R.K. Gupta, born 1966, is an Associate Professor in the Department of Zoology and Aquaculture, CCS Haryana Agricultural University, Hisar. He did his M.Sc. (1988) and Ph.D. (1992) in Zoology both from, CCS Haryana Agricultural University, Hisar. He has achieved many laurels in his cap right from doing PG diploma in agriculture journalism to member of various academic bodies associated with upliftment and dissemination of education in India. He has remarkable credit of holding several organizational memberships, attending and actively participating a number of refresher courses, training programmes, workshops, seminars, symposia and conferences at national and international levels on different aspects of science, thus gathering vast knowledge of current developments. He has about 100 publications to his credit in various journals of national and international reputes.

Dr. N.K. Yadava, born on May 18, 1950, is working as Professor in the Department of Zoology and Aquaculture, CCS Haryana Agricultural University, Hisar. He did his M.Sc. (1973) in Zoology from Gujarat University, Ahmedabad with a specialization in fisheries in fresh as well as marine water and Ph.D. (1987) in Zoology with a specialization in fresh water fisheries from CCS Haryana Agricultural University, Hisar. He has in total an experience of 31 years of teaching, research and extension in the fields of aquaculture and fisheries with special reference to fish ecology, reproduction, aquatic pollution management, utilization of livestock wastes for the production of fish through low cost inputs and fish pathology and management. He has guided a number of M.Sc. and Ph.D. students in the above research fields. He has published about 100 research papers including semi technical articles in various national and international journals of repute. He has carried out training cum-collaborative post-doctoral studies/research and teaching programme in the field of fish pathology and nutrition at the institute of aquaculture, Stirling University, Stirling, Scotland, UK in 1998 and visited many foreign universities/fisheries institutes such as University of Malaya, Malaysia; Fisheries Marine Laboratory, Philippines; University of Singapore; University of London; Cambridge University and Reading University, UK.

Dr. K.L. Jain, born on January, 1949, is an experienced teacher and researcher. He has, in total, an experience of 35 years of teaching and research in the fields of Bee Biology and behaviour, Environmental ecology and Animal toxicology and has guided a number of PG students in these fields. He did his Ph.D. as ICAR Senior Research Fellow in 1985 in the field of Insect Gustatory Physiology. He has published about 75 articles in various national and international journals of repute. He has been a life member of many academic societies of his research related fields and has attended a number of national and international conferences and symposia in India. He has also to his credit the coveted ICAR award "The Rafi Ahmed Kidwai Memorial Prize" for 1986 in the field of Agricultural Entomology. He worked as Head of the Department of Zoology and Aquaculture in CCS Haryana Agricultural University, Hisar for four years on a tenure post during 2000–2004. Presently he is working as Professor in the Department of Zoology and Aquaculture.

Dr. R.C. Sihag was born, 5 January, 1952. He is an experienced teacher, researcher and administrator. He has over 26 years of experience in teaching, research and extension in the fields of Bee Behaviour and Ecology and Environmental Biology and has guided a number of PG students in these fields. He has published about 150 articles in various national and international journals of repute. He has been a life member of many academic societies of his research related fields and has attended and chaired sessions in a number of national and international conferences and symposia in India and abroad. He has also to his credit the coveted ICAR award "The Rafi Ahmed Kidwai Memorial Prize" for 1986 in the field of Agricultural Entomology. He worked as Head of the Department of Zoology and Aquaculture in CCS Haryana Agricultural University, Hisar for three years on a tenure post. Presently he is working as Dean, College of Basic Sciences and Humanities, CCSHAU, Hisar.

Dr. V.P. Sabhlok was born on May, 1953. He is an experienced teacher and researcher. He has in total an experience of 27 years of teaching and research in the fields of Reproductive biology and nutrition. Presently he is working as Professor in the Department of Zoology and Aquaculture, CCS Haryana Agricultural University, Hisar. He has published about 50 research articles in various journals and books.